World Hunger Series 2007

Hunger and Health

A United Nations World Food Programme publication, Rome

Map A – Hunger and health across the world

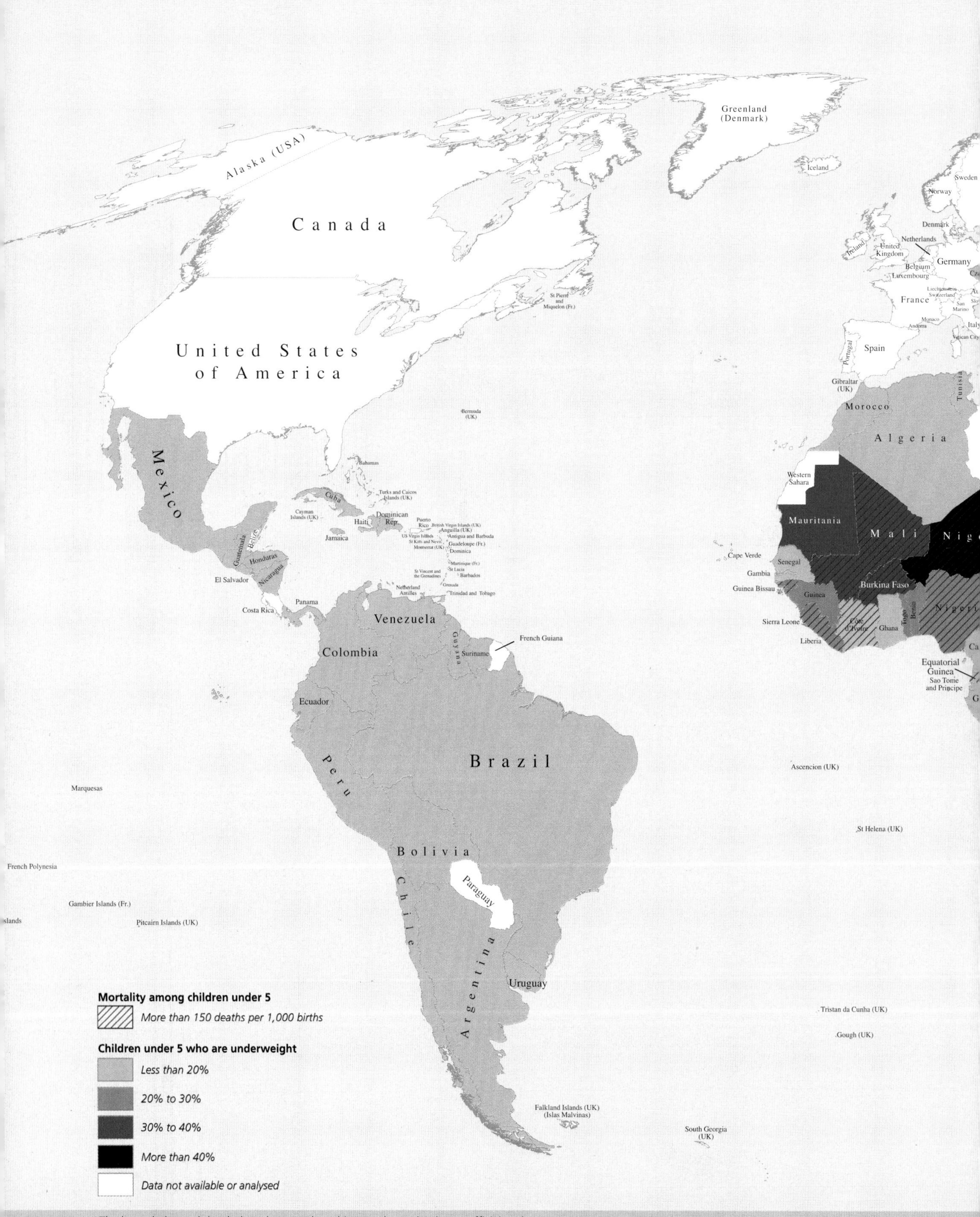

The boundaries and the designations used on this map do not imply any official endorsement or acceptance by the United Nations. Map produced by WFP VAM.

Data source: WHO, 2007

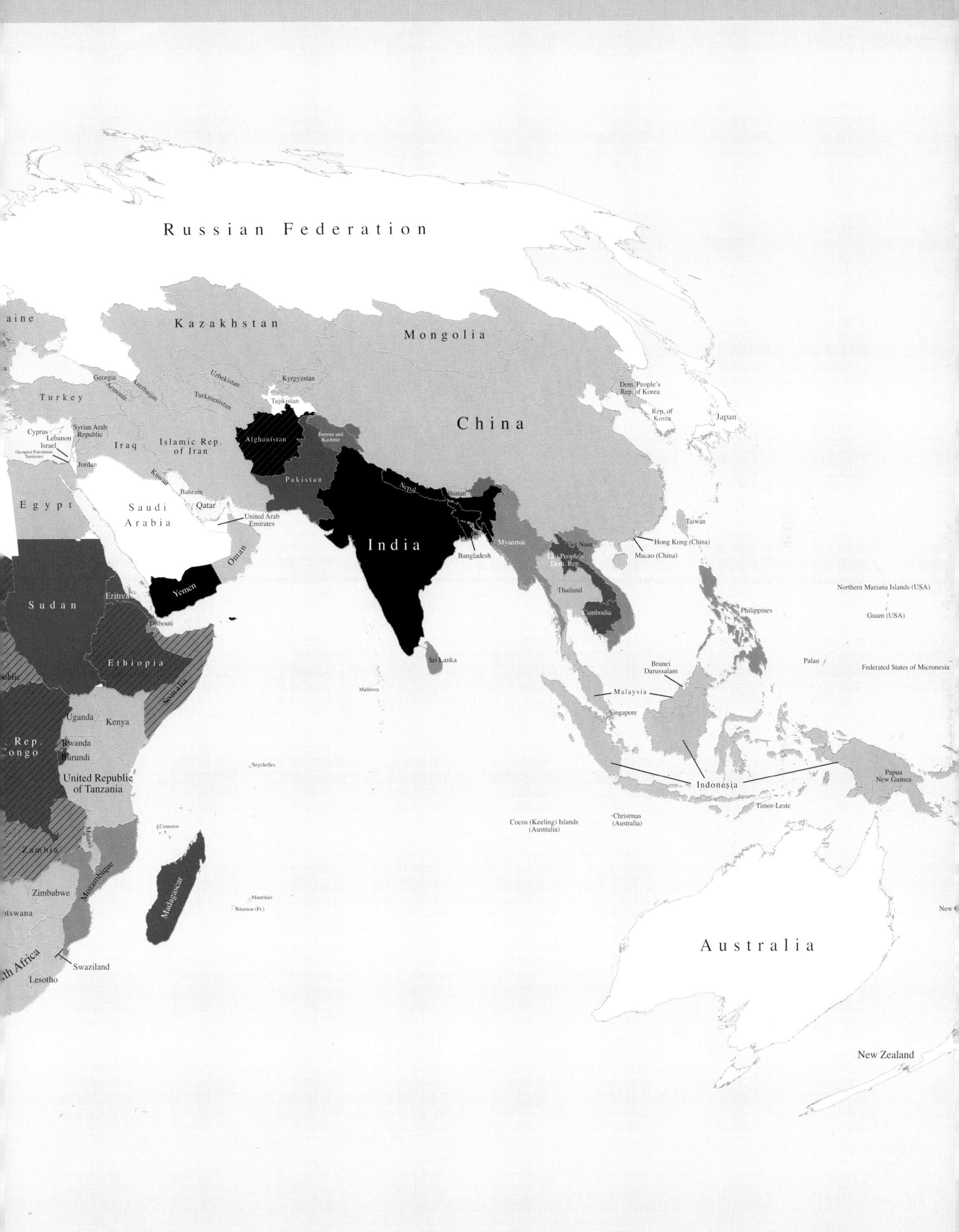

wfp vam
vulnerability analysis and mapping

Founded in 1963, WFP is the world's largest humanitarian organization and the United Nations' frontline agency in the fight against global hunger. WFP uses food assistance to meet emergency needs and support economic and social development.

Operational in 77 countries, WFP relies exclusively on donations. In close collaboration with other members of the United Nations family, governments and non-governmental organizations, WFP works to put hunger at the centre of the international agenda, promoting policies, strategies and operations that directly benefit the hungry poor.

Many people in the United Nations World Food Programme (WFP) were instrumental in preparing this latest report in the *World Hunger Series: Hunger and Health*.

Stanlake Samkange, Director of the Policy, Strategy and Programme Support Division, guided the effort. Deborah Hines was the lead author and team leader. The *World Hunger Series* team responsible for data collection and analysis included Federica Carfagna, Bruce Crawshaw, Peter Gray, Tomoko Horii, Rebecca Lamade, Kartini Oppusunggu, Robert Palmer, Livia Paoluzzi and Elena Vuolo – all made valuable contributions to the report.

Many experts generously made important technical contributions. In particular, WFP would like to thank: Steve Collins, Lorraine Cordeiro, Camila Corvalan, Paul Farmer, Wafaie Fawzi, Fernando Fernholz, Rosemary Fernholz, Stuart Gillespie, Carlos Guevara Man, Jean-Pierre Habicht, Alistar Hallam, Joan Holmes, Channa Jayasekera, Gina Kennedy, Richard Longhurst, Nkosinathi Mbuya, Saurabh Mehta, Gretel Pelto, Fabiola Pueda, Kate Sadler, Anna Taylor, Andrew Thorne-Lyman, Melody Tondeur, Patrick Webb and Stanley Zlotkin.

WFP would like to acknowledge the contributions of the United Nations World Health Organization (WHO) in providing updated data and technical support. Specifically, WFP extends its appreciation to Zoe Brillantes, Valentina Buj, Quazi Monirul Islam, Dermot Maher, Aayid Munim, Mercedes de Onis, Tikki Elka Pang, Randa Saadeh, Akihiro Seita and Kenji Shibuya.

Several people provided helpful comments and support: Carlos Acosta, Claudia AhPoe, Robert Black, Martin Bloem, Henk-Jan Brinkman, Marco Cavalcante, Gyorgy Dallos, Agnes Dhur, Francisco Espejo, Kul Gautam, Aulo Gelli, Ugo Gentilini, Paul Howe, Michael Hutak, Robin Jackson, Dan Lewis, Saskia de Pee, Roman Rollnick, Joseph Scalise, Anne Strauss, Judith Thimke, Tina Van Den Briel and Steven Were Omamo.

Special thanks is extended to WFP's Vulnerablility Analysis and Mapping Unit for the design and construction of the maps: Joyce Luma, George Mu'Ammar and Paola de Salvo.

Production of the report was supported by Cristina Ascone, Caroline Hurford, Francis Mwanza and Anthea Webb of the WFP Communications Division, and by Mark Menhinick and Marie-Françoise Perez Simon of the Translation and Documents Unit.

The battle against hunger can be won in our lifetime. The technology, knowledge and resources exist to meet the needs of the world's hungry. What is required now is for leaders to make the right political choices to ensure that images of hungry children are a problem of the past, rather than a shame of the present.

This edition of the *World Hunger Series* focuses on one of the most critical choices: taking action to address hunger and health together. For those in poor health, overcoming hunger is often a prerequisite for treatment and recovery. Food helps to speed recovery and guard against infection.

The 2007 *World Hunger Series* identifies proven solutions to ensure that research, policy and programmes reduce hunger and poor health for all people. For example, it shows that a combination of integrated food and health-related interventions is often better than a single disease approach. It also suggests that there are critical junctures when the benefits of reducing hunger and improving poor health have a particularly long-term impact. For instance, there is growing evidence to show that when pregnant women and especially adolescent girls in their first pregnancy are hungry, the well-being of future generations is jeopardized.

The report also makes the point that a greater alignment of efforts is necessary to address hunger and poor health effectively. National frameworks, policies, institutional arrangements, capacity-building and research all need to work in tandem as part of a coherent strategy, helping countries to be successful in creating a hunger-free population that is healthier, more productive and better able to learn.

Importantly, these actions make economic sense. The solutions are cost-effective and have long-term benefits for individuals, families, communities and nations. However, we must act not only for economic reasons – ending hunger is a moral imperative. The choices are before us. Leaders need to make the correct choices today, so that future generations will not suffer from hunger.

Josette Sheeran

Josette Sheeran
Executive Director
United Nations World Food Programme

Contents

About the United Nations World Food Programme 4

Acknowledgements 5

Foreword 6

Introductory note 10

Preface 11

OVERVIEW 12

PART I THE GLOBAL HUNGER AND HEALTH SITUATION 17
Introduction 19
1.1 Hunger, health and well-being 21
Intermezzo 1: An overview of micronutrient deficiences 24
1.2 Who are the most vulnerable? 33
Intermezzo 2: Women and ending hunger – the inextricable link 40
1.3 Tracking the hunger and health MDGs 43
1.4 Accelerating progress: making the right choices 53
Intermezzo 3: Hunger and disease in crisis situations 56

PART II UNDERNUTRITION AND DISEASE: IMPACTS THROUGHOUT THE LIFE CYCLE 59
2.1 Undernutrition and disease: a close relationship 61
2.2 A closer look at undernutrition and disease 64
Intermezzo 4: AIDS and hunger – challenges and responses 74
Intermezzo 5: Food support and the treatment of tuberculosis 77
Intermezzo 6: Nutrition transition in Latin America – the experience of the Chilean National Nursery School Council Programme 80
2.3 Emerging threats 82

PART III NATIONAL DEVELOPMENT: COMMITMENT AND POLITICAL CHOICE 87
3.1 Hunger impacts on human development 89
3.2 Effective solutions 92
Intermezzo 7: Sprinkles – an innovative, cost-effective approach to providing micronutrients to children 101
3.3 Making the right political choices 104
Intermezzo 8: Partnerships to overcome child undernutrition in Latin America and the Caribbean . . . 110
Intermezzo 9: From research to action 114

PART IV THE WAY FORWARD: TOWARDS A WORLD WITHOUT HUNGER 117
4.1 The way forward: ten key actions 119
Intermezzo 10: Nutrition a priority in Thailand 124

PART V RESOURCE COMPENDIUM 129
Overview: Technical notes 131
Table 1 – What does a hungry world look like? 133
Table 2 – How many people suffer from hunger throughout the life cycle? 137
Table 3 – What does a world with poor health look like? 141
Table 4 – How many people suffer from poor health during their life? 145

Table 5 – Who is at risk of hunger and poor health in crisis situations? . . . 149
Table 6a – How many people suffer from hidden hunger and childhood diseases? . . . 153
Table 6b – How many people suffer from infectious diseases? . . . 157
Table 7 – How many people are affected by natural disasters? . . . 161
Table 8 – What solutions are available for addressing hunger and poor health? . . . 165
Table 9 – What resources are allocated to reduce hunger and poor health? . . . 169
Table 10 – Progress towards achieving the MDGs by 2015 . . . 173

PART VI ANNEXES . . . 179
Abbreviations and acronyms . . . 181
Glossary . . . 182
Bibliography . . . 186
Text notes . . . 198
Costing the essential solutions . . . 199
Methodology for maps . . . 202

Figures

Figure 1 – Transition of average stature and life expectancy . . . 21
Figure 2 – Per capita daily consumption . . . 23
Figure 3 – Number of health professionals in developing and transition countries . . . 27
Figure 4 – Access to health services in developing and transition countries . . . 27
Figure 5 – Hunger and health determinants . . . 28
Figure 6 – Natural disasters worldwide . . . 36
Figure 7 – Prevalence of child underweight in LIFDCs by region . . . 45
Figure 8 – Prevalence of undernourishment in LIFDCs by region . . . 46
Figure 9 – Hunger indicators by region . . . 47
Figure 10 – Progress on underweight: MDG 1 for developing LIFDCs . . . 50
Figure 11 – Progress on undernourishment: MDG 1 for developing LIFDCs . . . 50
Figure 12 – Progress on child mortality: MDG 4 for developing and transition LIFDCs . . . 51
Figure 13 – Progress on maternal mortality: MDG 5 for developing and transition LIFDCs . . . 51
Figure 14 – GDP, child mortality and underweight in developing and transition countries . . . 53
Figure 15 – Undernutrition and disease determinants . . . 62
Figure 16 – Major causes of death among children under 5 worldwide . . . 64
Figure 17 – The two-way relationship: malaria and micronutrient deficiencies . . . 68
Figure 18 – Estimated burden of TB . . . 73
Figure 19 – Malnutrition in children under 5 in LIFDCs . . . 79
Figure 20 – Historic relationship between grain prices and food aid volumes . . . 83
Figure 21 – Number of slum dwellers . . . 85
Figure 22 – Undernutrition and lifetime loss in individual productivity . . . 90
Figure 23 – Practical solutions for all stages of the life cycle . . . 92
Figure 24 – National samples of growth faltering . . . 97
Figure 25 – Practical solutions for pregnant and lactating women, infants and young children . . . 98
Figure 26 – Practical solutions for school-age children and adolescents . . . 99
Figure 27 – Government spending on health as a percentage of GDP . . . 106
Figure 28 – Government expenditure on health . . . 107
Figure 29 – ODA and food aid . . . 113

Maps

Map A – Hunger and health across the world 2
Map 1 – Hidden hunger across the world 30
Map 2 – Hunger and natural disasters 39
Map 3 – Inequality of hunger across the world 55
Map 4 – Mortality and childhood diseases 66
Map 5 – The burden of malaria across the world 70
Map 6 – HIV/AIDS mortality in children under 5 72
Map 7 – Health inequalities across the world 103
Map 8 – National commitment to health 126
Map B – Hunger and health across the world 204

Introductory note

Hunger and Health ***not only surveys current knowledge about the link between poor nutrition and health, but also details the mechanisms by which hunger saps health and destroys the promise of decent, long, and meaningful lives.***

Moving between Harvard and Haiti, the affluent and poor worlds in general, one learns many things about what is in truth one world. This is our world: Coca-Cola is often readily available for the poor, who have almost nothing nutritive to eat yet are afflicted by diabetes because they consume too many of the wrong nutrients; cellular telephones reach into the poorest corners of the world, places in which childbirth is fraught with lethal peril; an art exercise with Rwandan orphans reveals that, although many of them do not attend school and are unsure of the provenance of their next meal, they do know how to draw uncannily accurate renditions of American rap stars.

This is the world I inhabit as a physician working in Rwanda and Lesotho and Haiti and Boston; it's also, if less transparently, the world inhabited by those who will read a report about hunger; it's the world described, in unflinching terms, in a new and important United Nations World Food Programme report on *Hunger and Health*.

This report is of the utmost importance, as all those working among the poor know. Sound approaches are laid out in careful detail in *Hunger and Health*, which offers a concise prescription for food and nutrition security, a prescription buttressed by solid research and long experience. It is our great privilege to find, in the World Food Programme, an ally in the struggle for equitable access to food, which is part of equitable access to good health.

Hunger and Health draws on decades of pragmatic experience in alleviating "food emergencies" and seeking to break the cycle of poverty and disease, and provides sound policy recommendations for nations and international standard-setting bodies seeking to meet the Millennium Development Goals.

We are deeply in debt to those who have written and contributed to *Hunger and Health*. Let this report, and written commitments to fair trade, land reform and improved agricultural practices, serve as the roadmap that we must all follow to make hunger in the 21st century be seen, first, as obscene and, second, as a global sickness for which we have, already, the cure.

Paul Farmer, MD
Harvard Medical School and Partners In Health

"We are made wise not by the recollection of our past, but by the responsibility for our future."

George Bernard Shaw (1856–1950)

As we pass the half-way mark for meeting the Millennium Development Goals (MDGs), hunger and health issues are receiving greater attention than ever before through the actions, campaigns and investments in support of the MDGs. However, progress is uneven in realizing most of the MDG targets and gaps are still widening in some countries.

One particular gap relates to the interaction between hunger and poor health. Women and children are particularly affected by lack of access to quality food and health services. Mothers struggle to prevent hunger and illness, with the effects playing out from one generation to the next. Obstacles also remain to putting knowledge and experience into practice at the community, national and international levels.

Learning from the past and applying our shared history is often idealized, yet in practice, political realities may dictate that we start anew and approach the future optimistically, neglecting the lessons of the past. Even so, there is still time to apply our accumulated experience, learning and will to define practical strategies and programmes to eliminate hunger. Opportunities to capitalize on the synergistic relationship between access to quality food and healthcare can be seized. In order to accelerate progress and achieve the MDGs, scientific knowledge must be translated into action, good intentions and international conventions given substance, and decisions made to put available resources to the best use.

This report, the second in the *World Hunger Series* after the inaugural report on learning in 2006, aims to contribute to improved understanding of the relationship between hunger and health. This 2007 edition uses evidence-based experiences to highlight lessons from past development practices and lays out possible solutions to eliminate hunger.

The 2007 report forms an integrated part of the *World Hunger Series* with evidence-based analysis intended to inform policy, programming and advocacy, and is to be followed by reports on markets, crises and social exclusion. The *World Hunger Series* complements ongoing efforts by governments, the private sector and local actors, and encourages sound policies in support of sustainable and cost-effective solutions that will, it is hoped, allow governments to surpass the MDG hunger target set for 2015 and eliminate hunger in the coming decades. The report provides sufficient evidence to confirm that hunger and poor health are solvable problems; we only need to mobilize our collective knowledge and make the right choices to end hunger.

Lessons from the past can be swept aside with surprising ease. Equally, indifference and inaction can be replaced by concrete efforts that galvanize all actors to work together to eliminate hunger.

Hunger and poor health not equally shared

Over the past 50 years the world has witnessed unprecedented gains in hunger reduction and health. Globally, there has been a significant decline in child undernutrition and infant mortality. Many physical aspects of health have improved substantially: people are living longer and globally experiencing lower levels of poor health in childhood and early adulthood.

Hunger and poor health are not shared equally among all people; the burden falls largely on the marginalized poor, with further disparity by gender, age and ethnicity. The life cycle impacts of hunger and poor health can be profound when compromised health spans generations.

We see major differences between rich and poor countries. In the poorest, most food-insecure countries – low-income food-deficit countries – life expectancy for men and women remains less than 50 years as a result of prolonged food shortages, disease, conflict and unequal access to quality healthcare. Approximately half of all deaths in children under 5 are directly due to hunger.

Also, while there are improvements in reducing hunger in some countries and for selected groups, the global attainment of MDG 1 target 2 (see box on page 44) is not on track. In some parts of the world past progress is being eroded and sustainable solutions are still far off for the hungry. Progress towards meeting health-related MDGs, like those for hunger, is also uneven and well-off countries are improving their health at a faster rate than those that are worse-off.

Hunger and health: a close relationship

The *World Hunger Series 2007* explores the multiple relationships between hunger and poor health and how they affect the growth of individuals, physiologically and psychologically, and constrain the development of nations both socially and economically.

Hunger and poor health are strongly related to political and economic choices, which in turn reflect the priorities attached to budget allocations, quality of social services and community values. People who suffer from hunger in any of its forms are not the decision-makers, nor are they necessarily well represented by them.

Just as hunger and health are closely related, so are undernutrition and disease; the relationship between undernutrition and disease is bidirectional and mutually reinforcing. Undernutrition leads to a state of poor health that puts the individual at risk of infectious and chronic disease. Hungry people are much less effective in fighting disease than well-fed people. An undernourished child tends to suffer more days of sickness than a well-nourished child as undernutrition contributes directly to disease by depressing the immune system and allowing pathogens to colonize, further depleting the body of essential nutrients.

Infections, no matter how mild, have adverse effects on nutritional status. Further, acute and chronic infections can have serious impacts on nutritional status, triggering different reactions, including reduced appetite and impaired nutrient absorption. Even when nutrients are absorbed, they may still be lost as a result of the infection.

Solutions are known and cost-effective

Despite the broad acceptance of the causal relationship between undernutrition and disease, resources have disproportionately been directed toward managing infectious diseases rather than preventing hunger and undernutrition. It is imperative that national frameworks and programmes are designed to consider the relationship between hunger and poor health. Only by prioritizing the hungry – and especially women and children at all stages of the life cycle – and by supporting principles of inclusion, equality, ease of access and transparency can the hungry benefit from the technological innovations that are transforming the world.

Reducing hunger increases productivity by improving work capacity, learning and cognitive development, and health by reducing the impact of disease and premature mortality. Hunger and poor health directly affect human and social capital formation and economic growth. These effects are long-lasting and inter-generational, with impacts impeding the achievement of other global social goals.

For the first time in history, the world can direct enormous resources to overcoming hunger and poor health. There is growing recognition that the cost of inaction is high, both in economic and moral terms – and that the cost of action is modest by comparison. A number of proven solutions are available and affordable, but they have to be scaled up to reach the world's vulnerable and marginalized people. An enabling environment to convert knowledge into feasible action and to remove institutional blockages is essential; otherwise, it will be difficult to maximize the potential gains from growing public and private resources to tackle hunger and poor health.

This *World Hunger Series 2007* puts forth a package of proven, practical and cost-effective solutions to address the interrelated causes of hunger and poor health. These solutions, combining food-based activities with basic healthcare and prevention activities, form "essential solutions" for hunger and poor-health reduction. With an emphasis on impact throughout the life cycle, these essential solutions aim to prevent hunger and improve the health of hungry people and contribute to achieving the MDGs. They specifically aim to expand programmes aligned with two broad "windows of opportunity" – critical times in an individual's life: early life, focusing on mothers, infants and young children, and adolescence, which includes school-age children.

The proposed essential solutions emphasize addressing common underlying factors, combining effectively the resources and tools at hand (including food and non-food resources), and scaling up what works. If programmes are built around the linkages between hunger and health, they will better address interrelated problems in a more holistic way.

They also highlight that general improvements in dietary intake, through improved access to quality food, in particular for young children, are likely to have a large impact on reducing the burden of disease.

Broadening commitment

Despite the various cost-effective solutions for combating hunger and improving health, and the potential to direct national and international political commitments to address these related problems for the poorest people, efforts are still insufficient. There is a real risk that the MDGs, themselves relatively modest, may not be met. The *World Hunger Series* challenges leaders to build on past successes, combining current knowledge with a will to undertake practical and effective solutions to end hunger in the coming decades.

There are four strong motivations for prioritizing these hunger–health solutions:

- The cost of hunger and poor health is high.
- Solutions are affordable, cost-effective and sustainable.
- There is consensus on the human right to adequate food, nutrition and health for all.
- Well-fed and healthy populations contribute to economic growth more effectively.

In the end, commitment determines whether interventions are effective and sustainable. The elimination of hunger cannot be relegated as a subsidiary goal of other commitments. In view of the tremendous human, economic and social costs of hunger, its elimination must be a development priority and an integral part of health goals.

To achieve optimal impact, appropriate resources are essential and their use must be maximized. The resources needed are not only financial: they include leadership, management and system support to make social services effective. To scale up activities, it is important to measure results and to know what works. Subsequently, resources can be allocated to projects that achieve impact:

- There is increasing evidence that nutrition and food support accompanying treatment for tuberculosis, human immunodeficiency virus and other infectious diseases increases adherence and improves outcomes, particularly for the poor. This support should become an integral part of treatment programmes. Research should be accelerated to improve the effectiveness of food and nutrition support aligned with treatment.
- The pervasive problem of micronutrient deficiencies shows that calories alone are not sufficient for good health. There is a need for increased awareness and understanding with regard to the value of micronutrients throughout the life cycle.
- Food fortification occurs in a number of countries, but more needs to be done. Multiple-micronutrient fortification of commonly consumed products and/or supplements may be a cost-effective strategy to address multiple deficiencies among school-age children, adolescents, refugees and internally displaced people. Also, more consideration should be given to fortifying food in the household (home fortification).

Making the right decisions

Urgent action is needed if hunger is to be eradicated in the coming decades. Government commitment to surpassing the MDGs, eradicating hunger and providing access to quality healthcare for hungry and marginalized people is the only option. The burden of hunger and poor health and its effect on national development can be only part of the rationale for acting. Action must address the human suffering caused by hunger and poor health and remove the divide between those who have access to sufficient quality food and healthcare, and those who miss these most essential ingredients for equitable human well-being.

We need to mobilize our collective will to make the right choices. The cost of inaction is high – economically, politically and, most importantly, morally.

What is hunger?

Most people intuitively understand the physical sensation of being hungry. But specialists who work on hunger issues have developed a range of technical terms and concepts to help them better describe and address the problem. Unfortunately, there is some disagreement on what these terms mean and how they relate to each other. This box provides a short glossary of these terms and concepts as used in this report. It cannot claim to be the only "correct" usage, but it does offer a relatively clear and consistent way of understanding the issues.

HUNGER. A condition in which people lack the required nutrients, both macro (energy and protein) and micro (vitamins and minerals), for fully productive, active and healthy lives. Hunger can be a short-term phenomenon or a longer-term chronic problem. It can have a range of effects from mild to severe. It can result from people not taking in sufficient nutrients or their bodies not being able to absorb the required nutrients. It can also result from poor food and childcare practices.

MALNUTRITION. A physical condition in which people experience either nutritional deficiencies (undernutrition) or an excess of certain nutrients (overnutrition).

UNDERNUTRITION. The physical manifestation of hunger that results from serious deficiencies in one or a number of macronutrients and micronutrients. The deficiencies impair a person from maintaining adequate bodily processes, such as growth, pregnancy, lactation, physical work, cognitive function and resisting and recovering from disease.

UNDERNOURISHMENT. The condition of people whose dietary energy consumption is continuously below a minimum requirement for fully productive, active and healthy lives. It is determined using a proxy indicator that estimates whether the food available in a country is sufficient to meet the energy (but not necessarily the protein, vitamins and minerals) requirements of the population. Unlike undernutrition, the indicator does not measure an actual outcome.

SHORT-TERM HUNGER. A transitory form of hunger, including "hunger pangs", that can affect short-term physical and mental capacity.

FOOD SECURITY. A condition that exists when all people at all times are free from hunger. The concept of food security provides insights into the causes of hunger. Food security has four parts:

- availability (the supply of food in an area);
- access (a household's ability to obtain that food);
- utilization (a person's ability to select, take-in and absorb the nutrients in the food); and
- vulnerability (the physical, environmental, economic, social and health risks that may affect availability, access and use) (WFP, 2002; Webb and Rogers, 2003).

Food insecurity, or the absence of food security, is a state that implies either hunger resulting from problems with availability, access and use or vulnerability to hunger in the future.

What is the difference between hunger and undernutrition?
Undernutrition is the physical manifestation of hunger. It can be measured using indicators such as:

- weight-for-age (underweight);
- height-for-age (stunting); and
- weight-for-height (wasting).

In some cases, undernutrition can be caused by disease, which influences the adequacy of food intake and/or its absorption in the body (and therefore the level of hunger). Disease affects the adequacy of food intake by altering metabolism (thus increasing the requirements for the intake of nutrients) and reducing appetite (often reducing the amount of food ingested). At the same time, disease may cause problems of absorption through the loss of nutrients (e.g. vomiting, diarrhoea) or its interference with the body's mechanisms for absorbing them. Thus disease aggravates undernutrition. Of course, disease often has many other serious and debilitating effects not directly related to its impact on hunger.

How is hunger related to undernutrition and food insecurity?
Hunger, undernutrition and food insecurity are nested concepts. Undernutrition is a subset of hunger, which in turn is a subset of food insecurity (see diagram below).

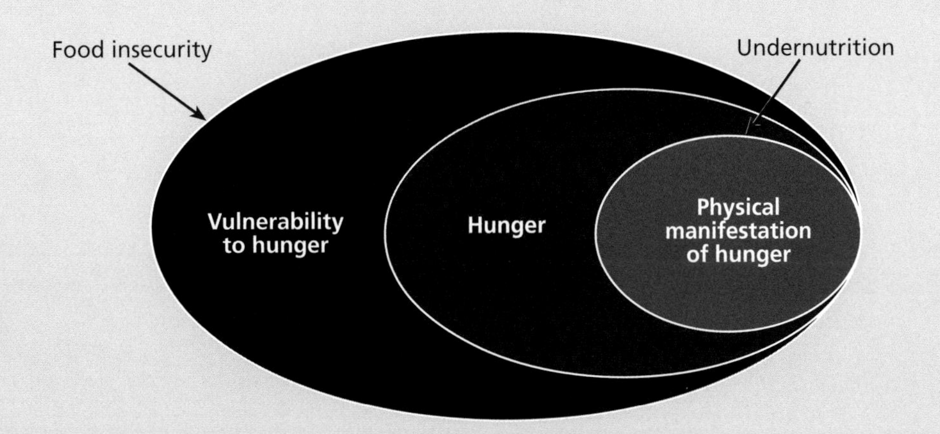

PART I The Global Hunger and Health Situation

Hunger and health are inherently related. Health cannot be improved without tackling the problem of hunger; hunger in turn leads to poor health, and many of the causes of hunger also contribute to poor health.

Part I reviews the current hunger and health situation and tracks progress towards meeting the Millennium Development Goals. **Chapter 1** lays out the bidirectional relationship between hunger and poor health, showing that it is difficult to significantly improve the health of an individual without eliminating all forms of hunger. **Chapter 2** shows what a hungry world and a world in poor health looks like and profiles the groups most vulnerable to hunger and poor health. It presents the development challenges before us, examining the multiple causes of hunger and poor health and the devastating role of conflict and natural disasters in impeding progress in hunger reduction. **Chapter 3** provides an update on progress towards meeting the MDGs for hunger and health, showing that progress is still not sufficient. **Chapter 4** lays out some choices that leaders may make to accelerate progress towards meeting the MDGs.

Introduction

"Why does hunger persist in a world of plenty? One of the greatest questions of our time, this is also a question of earlier times. ... the history of hunger is embedded in the history of plenty."

Sara Millman and Robert W. Kates, 1990

Historical social and economic transitions illustrate interesting patterns of progress and regress. From the 21st century perspective, it is easy to assume that transitions represent remarkable strides to feed the world and protect it against disease. Over the last 200 years there have been important gains in hunger reduction; however, 854 million hungry people throughout the world still struggle to survive and more than 16,000 children die needlessly each day from hunger-related conditions. The progression to a world free from hunger is uneven, and it is clear that progress in tackling hunger and related health issues is held back by significant obstacles.

A number of development models suggest that improved diet and nutrition lead to better health, which results in greater equity (Semba, 2001). However, each country faces a unique set of challenges regarding the extent and type of hunger and the causes of prevailing health problems. The interwoven causes of hunger and poor health are deeply rooted in social, economic and political conditions.

Hunger and poor health are thus strongly related to political and economic choices, which in turn reflect the priorities attached to budget allocations, quality of social services and community values. People who suffer from hunger in any of its forms are not the decision-makers, nor are they necessarily well represented by them.

The *World Hunger Series 2007* explores the multiple relationships between hunger and poor health and how they affect the growth of individuals, physiologically and psychologically, and constrain the development of nations both socially and economically.

There are numerous discussions on hunger. Too often, however, hunger analyses focus exclusively on the physical manifestations, portraying acutely undernourished children with swollen bellies or stunted children affected with inadequate growth at critical periods of their lives. The causes often focus on poor food production or insufficient income to purchase the quality food required by a household. These may be central to any discussion on hunger, but it is important to emphasize that there are knowledge gaps in the hunger debate. The *World Hunger Series* attempts to shed light on some of the neglected areas in the discussion, and in this edition specifically, the hunger–health relationship. This *World Hunger Series* examines the profound effect that hunger has on health, including disease prevention and treatment, and gives special attention to nutrient absorption and utilization.

This report also addresses programming issues and the health actions necessary to overcome hunger: how hunger and health interventions can be better aligned and strategic key actions implemented to limit the damaging impact of hunger on health and well-being.

The *World Hunger Series 2007* argues that the factors that jeopardize good health and reinforce hunger are generally well known, and that affordable solutions are widely available. The report presents a call to action that lauds the unprecedented global efforts to tackle hunger and poor health. It also highlights that opportunities abound for concentrating resources in collaborative, harmonized approaches that support national and local frameworks. Five messages underlie the *World Hunger Series 2007*:

- Hunger and poor health are related global problems.
- They disproportionately affect the poorest and most vulnerable, needlessly shortening lives and the quality of life for hundreds of millions.
- Women and children are particularly affected by hunger and poor health.
- Poor health and hunger impact national development, both now and for future generations.
- Hunger and poor health are solvable problems; however, current approaches do not always lead to solutions that are equitably accessible and sustainable.

Introduction

The *World Hunger Series 2006* publication – *Hunger and Learning* – introduced the premise that political choice is directly related to the persistence of hunger. The 2007 report looks in more depth at how political choices influence progress in reducing hunger and achieving good health, and how these choices often ignore processes of marginalization and the inequalities that reduce access to quality food and health services by those who are most vulnerable.

This report emphasizes the profound and mutually reinforcing benefits that investments in hunger elimination and health can achieve for individuals and nations. It offers an agenda for concrete action at the community, national and international levels. New alliances and partnership approaches are increasingly apparent in a globalizing world and these represent new opportunities for action. Mobilizing the various actors to work jointly in the same direction is vital and the *World Hunger Series* makes a modest contribution to that end. This *World Hunger Series 2007* report has five principal parts:

- **The global hunger and health situation** surveys the current state of hunger and poor health in the world.
- **Undernutrition and disease: impacts throughout the life cycle** explores the two-way relationship between hunger and health during the life cycle and identifies knowledge gaps that if addressed would enhance current hunger reduction efforts.
- **National development: commitment and political choice** presents the rationale for increasing commitment to fight hunger and poor health. It also examines the role of hunger reduction in health programmes and in national development. Furthermore, it presents evidence showing that cost-effective solutions are at hand and have contributed to positive health impacts.
- **The way forward: towards a world without hunger** sets out concrete actions for moving ahead with integrated, harmonized solutions within government frameworks.
- A **resource compendium** contains supporting data.

1.1 Hunger, health and well-being

"The causes of hunger have ranged from drought, flood, and other natural disasters that curtail food production, to hierarchical social structure with its rules of inclusion and exclusion that limit the distribution of food and wealth ..."

Lucile F. Newman et al., 1990

The basis of hunger

Since the beginning of human existence people have struggled to be free of hunger, undergoing periods of plenty and times of food scarcity, of growth and decline. From the Late Palaeolithic period – 30,000 to 9000 BC – hunters and gatherers persevered to manage the rich biodiversity of their environment, achieving an average height that is still not equalled today.

By the 3rd millennium BC, average stature had declined by 11 cm for men and 14 cm for women (Cohen and Armelagos, 1984). During this period of settlement and increased population, diets changed: people consumed a larger proportion of grains, drawing from a dramatically reduced number of plant species, and ate less meat (Barnes, 2007). Permanent settlements and the shift from hunting and gathering created possibilities for food storage and specialized roles in societies. However, they were accompanied by social stratification and hunger (Milton, 2000).

Figure 1 – Transition of average stature and life expectancy[1]

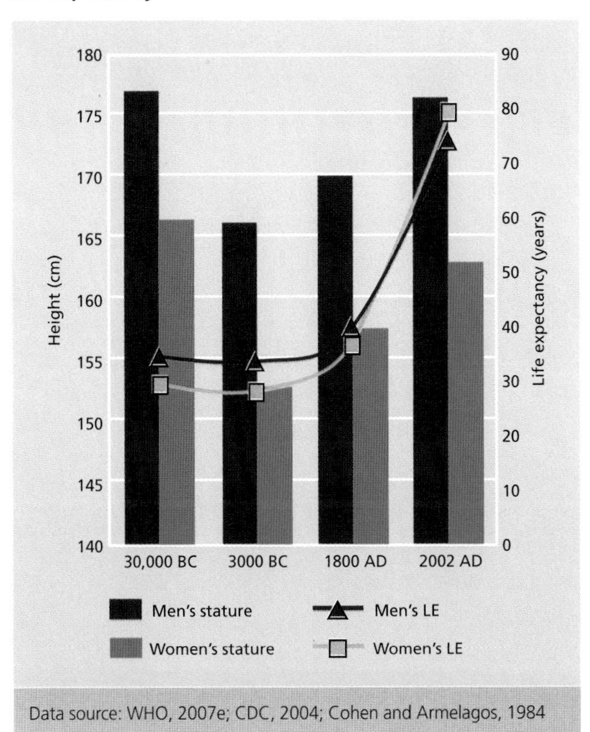

Data source: WHO, 2007e; CDC, 2004; Cohen and Armelagos, 1984

This shift from Palaeolithic to Neolithic is just one of the numerous transitions that changed diets and affected health over time. These transformations coincided with technological innovations, demographic transitions and social transformations; each brought progress as well as food shortages and hunger for certain parts of the population.

Another transition, the industrial revolution of the 1800s, created new wealth and improved nutritional status in a number of countries for the upper and middle classes, resulting in longer life expectancy (LE) and increased stature: life expectancy increased to 40 years; average stature was 157 cm for women and 170 cm for men (Cohen and Armelagos, 1984). But society also became more stratified and not all segments benefited from these improvements: in fact, the lower classes became further marginalized with the inequitable distribution of wealth, and hunger was pervasive.

We see similar patterns today with the "nutrition transition" – the concurrence of both overnutrition and undernutrition. We live in a world with immense wealth, huge disparities and unequal access to health services and food, even though globally there is sufficient food to feed all. In rich and healthy countries, men and women show a sustained increase in stature and life expectancy: life expectancy has increased to up to 75 years for men and 80 years for women in developed countries; average stature is 163 cm for women and 176 cm for men (WHO, 2007a; CDC, 2004).

But again we see major differences between rich and poor countries. In the poorest, most food-insecure countries – low-income food-deficit countries (LIFDCs) – life expectancy for men and women remains less than 50 years as a result of prolonged food shortages, disease, conflict and unequal access to quality healthcare. As well, the current nutrition transition combines greater access to poor quality food with increasingly high rates of overweight, alongside a continuing high prevalence of underweight. Importantly, foetal undernutrition increases the risk of overnutrition later in life. A survey in China found that one in five overweight children under the age of 9 had suffered from stunting as a result of chronic hunger early in life (World Bank, 2006).

What causes hunger

The causes of hunger have ranged from natural disasters that reduce food production to hierarchical social constructs that limit the distribution of food and wealth. Today, the hundreds of millions of people living with hunger face a combination of factors that leave their most basic food needs unmet and that perpetuate hunger from one generation to the next. These factors include changing agricultural patterns, food prices, incomes, seasonal food shortages, food preferences, cultural and gender dynamics, urbanization, conflict, natural disasters, changing climatic patterns, and political choices such as those that favour defence spending over allocations for health and food-based social safety nets. "World food production is more than enough to feed everyone on earth; yet hunger still reigns in most parts of the world and remains a constant threat to humanity" (Salleh, 2001).

About 10 percent of the world's hungry people endure acute food shortages as a result of crises. The plight of 2 million displaced people in Darfur in the Sudan is just one tragedy of the first decade of the new millennium. The other 90 percent of the world's hungry live with chronic hunger – nagging hunger that does not go away. The chronically hungry lack regular access to sufficient quality food and often suffer from poor health. Their basic means of livelihood progressively deteriorate, and eventually survival becomes the main concern. Many social and economic policies seem only to entrench hunger and poor health.

The resources allocated to chronic hunger are dwarfed by those for humanitarian operations. In 2006 alone WFP provided US$742 million in assistance for 15 countries facing the most dramatic situations of chronic hunger. In addition to this, WFP provided US$558 million for the Sudan in the same year, illustrating the immense toll that conflict takes on societies, the level of resources required to save lives in crisis situations and how little is left over to fight chronic hunger.

Stature as proxy for growth and development

Empirical evidence shows that human height is a proxy of well-being and health. Height among children can help to predict survival and is correlated with later school and labour market performance. A child's height by the age of 4 is a good predictor of adult height (Martorell and Habicht, 1986); thus adult height may be treated as predetermined from early childhood (Schultz, 2002).

Early studies on the relationship between height and socio-economic development were conducted by medical doctors and nutritionists such as Nevin S. Scrimshaw, who documented that children from Central America and Panama were shorter than children from the United States and argued: "... that the nutritional damage which the child suffers after weaning represents a very serious human and economic loss to the country ... in terms of health and resistance to disease and consequently in capacity for learning and working" (Scrimshaw et al., 1959).

The relationship between stature and productivity was first measured by G. B. Spurr, who studied the productivity of Colombian sugarcane cutters aged 18–34 and found that height, as a proxy for the history of nutritional status, made a positive contribution to productivity. Spurr's results indicated that undernutrition had an important negative effect on underdeveloped economies (Spurr et al., 1977).

Calories are not enough

"Hidden hunger" affects more than 2 billion people, even when they consume adequate calories and protein (United Nations Millennium Project – Hunger Task Force, 2005). Hidden hunger is present when people lack one single micronutrient alone, or a combination of micronutrients. Micronutrient deficiencies result in serious public health problems primarily because they compromise the immune system, allowing infections to take hold.

Mothers and children are among the most vulnerable and affected, although all may experience reduced productivity and socio-economic opportunities. As acknowledged by the nutrition community, hidden hunger "... is a hunger that does not manifest itself in the form of a bloated belly or emaciated body. But it strikes at the core of people's health and vitality" (Gautam, 2006).[2]

How many calories are enough?

Nobody can say precisely how many calories a particular person needs to survive, to be free from hunger or to live a healthy life. Various factors, starting with an individual's make-up, the harshness of their environment, activity level and general health, all coalesce to determine the appropriate energy intake for an individual. In a tropical country people generally require less energy than in a country with harsh winters, or those situated at high altitudes. Illness also affects caloric intake, increasing energy requirements on the one hand and decreasing appetite on the other.

Most experts agree that 2,100 kcal per day supplies sufficient energy for most people; it is a benchmark for average caloric consumption. But the average actual caloric consumption varies dramatically from country to country. A person in the United States has access to an average of 3,800 kcal per day. In Eritrea caloric consumption is about 1,520 kcal per day, less than half the United States consumption.

Put in another context, the standard ration for a refugee in a camp in Kenya is 2,100 kcal per day (WFP, 2007). Humanitarian agencies have moved from the 1970s policy of providing "survival" rations of 1,200–1,800 kcal to the 1980s "minimum standards" policy of 1,900 kcal to the target level of 2,100 kcal in the 1990s. The 2,100 kcal standard, while sometimes not met, provides a minimum threshold for sufficient calories.

Irrespective of the caloric threshold, it is clear what happens to a child or an adult who is not able to consume sufficient food and expends more calories than are consumed. The imbalance, if it continues, forces the body to draw on itself and consume its own tissues for energy, gradually allowing defences to drop and disease to take hold (Russell, 2005).

An anthropologist working among the Southern Bantu in the 1930s advanced an early understanding of hunger: "Hunger leads first to the concentration of the whole energy of the body to the problem of getting food. Every thought and emotion of the hungry is fixed on this one primary need. Failure to obtain sufficient foods gradually lowers the whole vitality of the body" (Richards, 2003).[3]

Figure 2 – Per capita daily consumption (kcal)

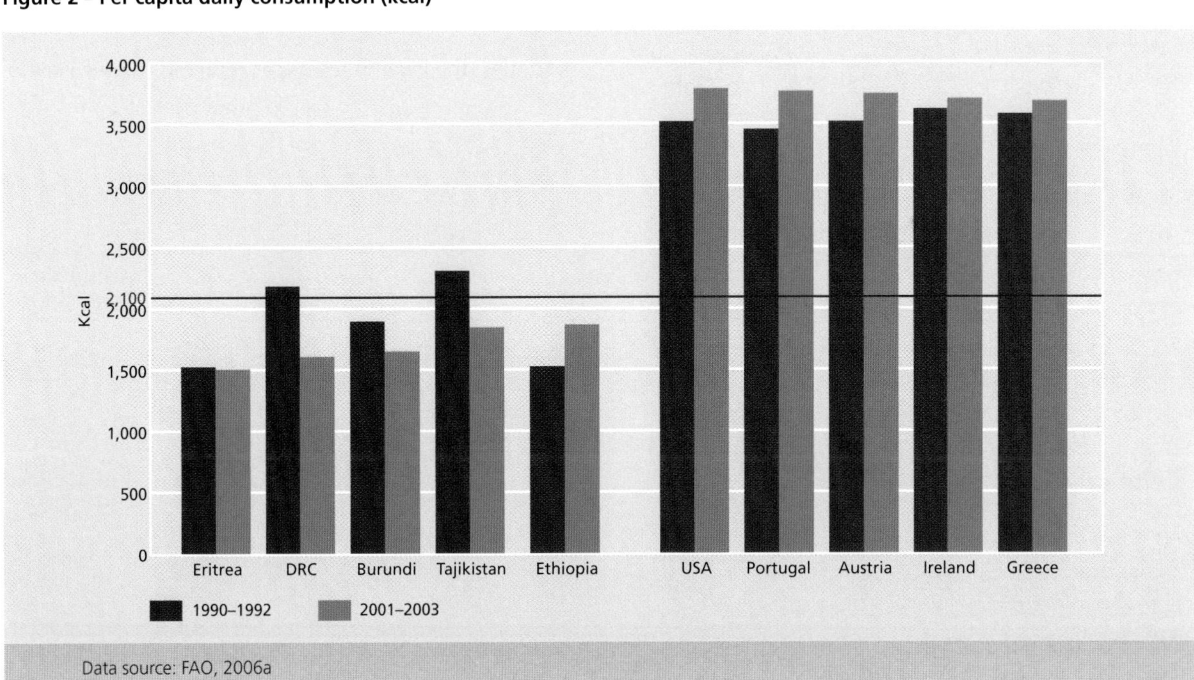

Data source: FAO, 2006a

Intermezzo 1: An overview of micronutrient deficiencies

The availability and absorption of micronutrients is one of the most important determinants of health. Hidden hunger, particularly deficiencies in vitamin A, iron, zinc and folic acid (folate), is the main contributor to the world's disease burden. Micronutrient deficiencies and disease form a vicious circle, because the immune system is weakened and it predisposes young children to infectious diseases. In turn, infection reduces appetite, decreases food intake and depletes the body of micronutrients, essential for adequate growth and development. The devastating effects of these deficiencies, for example blindness, anaemia, low birthweight, reduced cognitive development, and reduced productivity, are widely known.

Micronutrient deficiency	Burden of deficiency
Vitamin A deficiency	Nearly 800,000 deaths among women and children worldwide can be attributed to vitamin A deficiency (VAD). VAD is attributed to 20 percent of maternal deaths worldwide. South East Asia and Africa have the highest burden of VAD. Rice et al., 2004
Iron deficiency and anaemia	Iron deficiency contributes to 18.4 percent of total maternal deaths and 23.5 percent of perinatal deaths. Sanghvi et al., 2007
	Iron deficiency is attributed to 115,000 maternal deaths and 591,000 perinatal deaths globally. The total global burden attributed to iron deficiency anaemia is 814,000 deaths. Stoltzfus et al., 2004
Zinc deficiency	The estimated global prevalence of zinc deficiency is 31 percent. Zinc deficiency contributes to increased risk of childhood diseases, a main cause of death among children. Caulfield and Black, 2004
	It is estimated that 665,000 child deaths, or 5.5 percent, are related to zinc deficiency. Sanghvi et al., 2007
Folic acid deficiency	Access to adequate folic acid supplementation is estimated to reduce the incidence of neuronal tube defects – affecting up to 5 babies per 1,000 live births worldwide; 95 percent of cases occur from a first pregnancy. Gupta and Gupta, 2004

Vitamin A deficiency

Vitamin A is vital to healthy growth, especially for infants and young children. It regulates a number of biological processes, including growth, vision, reproduction and cellular differentiation. A person cannot survive without vitamin A. Vitamin A must be provided from the diet in adequate amounts to meet the body's physiological needs.

Recent trends indicate that there is a general decline in the prevalence of severe vitamin A deficiency; however, numerous studies have shown that mild vitamin A deficiency is pervasive in most of the developing world. It is the leading cause of preventable childhood blindness in developing countries. It is also an increasingly recognized problem among rural women in many countries and suspected to be a major underlying cause of maternal mortality. Newborns and women with vitamin A deficiency are at greater risk of illness. Immediate postnatal vitamin A supplementation in mothers leads to increased vitamin A passed on to the infant during breastfeeding (Basu et al., 2003).

Vitamin A deficiency

Life cycle stage	Hunger/health risk
Pregnancy	Implicated in maternal mortality.
Infants and young children	Increased risk of dying from diarrhoea, measles and other diseases. Increased risk of blindness, chronic ear infection and respiratory diseases.

Iron deficiency

Iron helps to produce energy by carrying oxygen to red blood cells. Iron deficiency is one of the most common nutritional disorders worldwide, stealing vitality from the young and the old and impairing the cognitive development of the undernourished.

The highest risk groups for iron deficiency are pre-term and low-birthweight babies, infants and children during periods of rapid growth, women of reproductive age, pre-menopausal women and pregnant women. It is important for women to enter into pregnancy with sufficient iron levels and maintain these throughout pregnancy; chronic iron deficiency can also lead to anaemia among lactating women (Dugdale, 2001).

Newborns and pre-school children face cognitive and cerebral damage with life-long negative effects caused by anaemia. Anaemia is also associated with hookworm infestation.

Iron deficiency anaemia

Life cycle stage	Hunger/health risk
Pregnancy	Increased risk of intra-uterine growth retardation, low birthweight, premature delivery and perinatal morbidity and mortality.
Infants and young children	Impaired motor development and coordination; impaired language development and scholastic achievement; psychological and behavioural effects; decreased physical activity; increased risk of acute respiratory infections.
School-age children and adolescents	Decreased learning capacity; increased risk of acute respiratory infections.
Adults and elderly people	Lower productivity and lethargy; increased risk of acute respiratory infections.

Folic acid deficiency

Folic acid is critical for pregnant mothers and newborn babies: it works with vitamin B12 to form healthy red blood cells. Folic acid helps to reduce the risk of neurological defects in foetuses and is essential for the development of an infant's neurological system (Green, 2002). If a mother remains underweight during pregnancy with a serious deficiency of folate concentration, the foetus faces an increased risk of pre-term delivery, low birthweight and growth retardation (Johnson et al., 2005).

Zinc deficiency

Most children in developing countries consume very small amounts of animal proteins – the dietary source of zinc with the highest bioavailability – which explains why zinc deficiency may be one of the most prevalent nutritional disorders in children in developing countries.

Zinc deficiency is largely related to inadequate intake or absorption of zinc from the diet, although excessive loss of zinc during diarrhoeal episodes is common. It is associated with difficulties during pregnancy and childbirth, compromised immune responses and increased risk of infectious diseases – zinc is a major factor in diarrhoeal disease, pneumonia and malaria, low birthweight and stunted child growth. Therefore, adequate intake of zinc becomes crucial for all children, as they generally suffer from a higher risk of infectious disease (WHO, 2002).

An expanded view of hunger goes beyond immediate caloric shortages or prolonged physiological hunger. It includes socio-economic aspects that consider food production and access, nutrient absorption and utilization, food practices and childcare practices. This view is based on more than 35 definitions or conceptualizations of hunger from applied research, with the aim of deepening our collective understanding of why people are hungry. It also incorporates an understanding of how the governments of the world can meet the challenge of ensuring that all citizens have access to sufficient quality food and a life free from hunger.

Food intake affects mood, behaviour and brain function, although the effects are difficult to quantify. A hungry person may feel irritable, restless, apathetic or moody over a longer period. Deficiencies of multiple nutrients, rather than one single nutrient, are responsible for changes in brain function that, if prolonged, can cause damage to the nerves in the brain.

Disorders related to mental health have dramatically increased among developing countries and are an emerging cause of morbidity. Children are more susceptible to shocks than adults because of their sensitive, developing neurological systems.

Shocks may include direct traumatic events, such as war or famine, or more subtle ones such as stunting or severe undernutrition leading to cognitive impairment (Margallo, 2005).

Directly linked to a life free from hunger is an individual's health and well-being (Holben, 2005).

Health and well-being

Over the past 50 years the world has witnessed unprecedented gains in health. Globally, there has been a significant decline in the infant mortality rate, the level of childhood immunization has increased dramatically and access to primary healthcare, including water and sanitation, continues to improve, albeit more slowly in the last decade than in previous decades. Many physical aspects of health have improved substantially: people are living longer and in general experiencing lower levels of poor health in childhood and early adulthood.

What is health?

WHO defines health as a state of complete physical, mental and social well-being, and not merely the absence of disease or infirmity. This complete well-being enables people to live socially and economically productive lives. This concept of health recognizes the unique challenges of achieving health in an inequitable and unequal world, and so seeks to place health in the context of both individual rights and global accountability for fulfilling those rights. In this expanded perspective, health incorporates both mental and physical aspects. These health-for-all values underpin all aspects of health policy and are based on four key health elements (Yach, 1998):

- recognition of the highest attainable standard of health as a fundamental right;
- continued and strengthened application of ethics to health policy, research and service provision;
- implementation of equity-orientated policies; and
- incorporation of a gender perspective into health policies and strategies.

Gains – but progress is uneven

Despite the rapid gains in health outcomes, progress is not equal, especially for the poor and marginalized. For example nearly 10 million children under 5 die every year, mainly from preventable diseases such as pneumonia and diarrhoea; and malaria kills a child somewhere in the world every 30 seconds, disproportionately in poor countries.

Fewer than 60 percent of children from sub-Saharan Africa receive vaccines against curable diseases (UNICEF, 2006b).

Surprisingly, maternal deaths have remained unacceptably high. Overall, in sub-Saharan Africa, 1 in every 16 women will die of pregnancy-related causes (UNFPA, 2005). Disparities in achieving conditions for safe motherhood and sufficient access to health services are most apparent in rural areas. For example,

Figure 3 – Number of health professionals in developing and transition countries (number per thousand people)

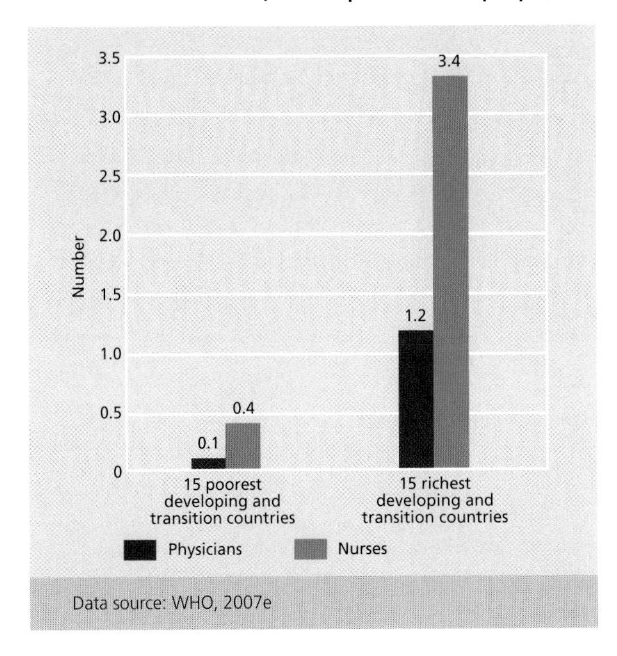

Data source: WHO, 2007e

Figure 4 – Access to health services in developing and transition countries

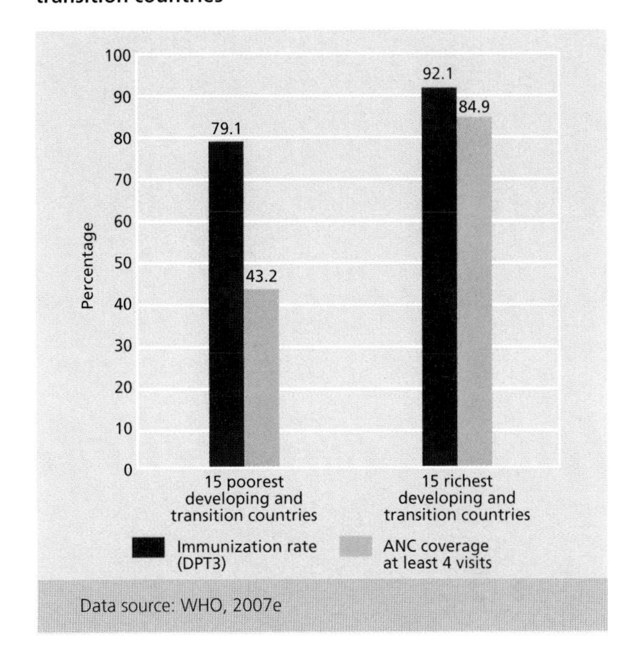

Data source: WHO, 2007e

pregnant women in urban settings are three times more likely to be assisted by skilled professionals than women from rural regions. This trend is particularly evident in sub-Saharan African countries, where the health professional drain is greater than in other regions of the world. Fewer than 5 percent of health professionals are in African countries; fewer than 15 percent are in South East Asian countries (WHO, 2006).

Challenges remain

Challenges remain, despite massive investments and impressive gains in health in the last decades. The hungry and poor tend not to be the main recipients of improvements. The correct policy choices must integrate health with those sectors that address hunger and other development constraints for the poor.

The hunger–health relationship

Hunger and health are inherently related: health cannot be improved without tackling the problem of hunger; hunger in turn leads to poor health. Many of the causes of hunger also contribute to poor health.

Hunger leads to low levels of energy, decreasing immunity and increasing susceptibility to poor health.

The two-way relationship of hunger and poor health is rooted in determinants that are layered (Schroeder, 2001). The most immediate causes of hunger are inadequate diet and illness; these are intimately linked with access to and utilization of food. Related are childcare practices, poor access to clean water and sanitation. In particular, a mother's poor health undermines the provision of adequate childcare. Finally, the underlying causes of hunger and poor health are influenced by the socio-economic and political environment.

Hunger increases the severity of infectious disease and therefore the risk of dying from a disease once ill. Individuals suffering from illness are unable to utilize nutrients properly and as a consequence are in a weakened state that compromises the immune response to infections.

Hunger and poor health also generate wide-ranging social and economic consequences and further entrench poverty and inequity. Hunger and poor health greatly reduce the ability to learn and work, diminish productivity and create dependency. Spiralling

Figure 5 – Hunger and health determinants

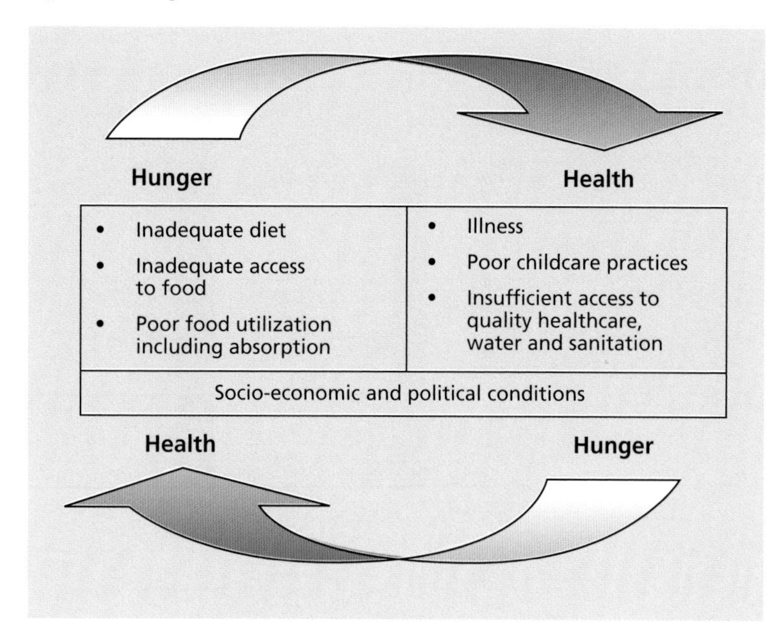

healthcare costs deplete family incomes and undermine the ability to purchase quality foods in sufficient quantities, degrading a poor family's standard of living. Household and village-level studies show that illness is a major shock to income earning and is one of the leading causes of households' decline into poverty (Krishna et al., 2004).

Hunger and poor health increase reliance on risky coping mechanisms. Insufficient access to food and mounting health bills quickly create an urgent need for cash. One coping mechanism is to reduce the quantity and quality of food consumption, either for the whole family or for selected members. The poor possess few liquid assets that can be used for such emergencies, so they may be forced to sell land or equipment central to sustaining their food supply and livelihoods. Another common coping strategy is to take children out of school and send them to work, depriving them of the skills needed to avoid a future of poverty (WFP, 2006b). Another is to undertake activities that increase the susceptibility to infectious diseases, for example sex trade.

When hunger and poor health afflict a family, young children are affected most; they will suffer to a greater degree from the combined effects of food shortages, reduced care and increased susceptibility to infectious diseases. Thus, understanding the relationship between hunger and poor health requires a long-term perspective; what happens at one stage of life affects later stages, and what happens in one generation affects the next.

Identifying the relationship between health and hunger throughout the life cycle is an aim of this report.

Map 1 – Hidden hunger across the world

The boundaries and the designations used on this map do not imply any official endorsement or acceptance by the United Nations.
Map produced by WFP VAM.

Data sources: WHO, 2007; The Micronutrient Initiative and UNICEF, 2004

Russian Federation
Kazakhstan
Mongolia
Kyrgyzstan
Uzbekistan
Turkmenistan
Tajikistan
Turkey
Syrian Arab Republic
Lebanon
Israel
Jordan
Iraq
Islamic Rep. of Iran
Afghanistan
Pakistan
China
Dem. People's Rep. of Korea
Rep. of Korea
Japan
Kuwait
Bahrain
Qatar
Saudi Arabia
United Arab Emirates
Oman
India
Bhutan
Bangladesh
Myanmar
Taiwan
Hong Kong (China)
Macao (China)
Thailand
Cambodia
Philippines
Northern Mariana Islands (USA)
Guam (USA)
Yemen
Eritrea
Djibouti
Ethiopia
Somalia
Sri Lanka
Maldives
Brunei Darussalam
Malaysia
Singapore
Palau
Federated States of Micronesia
Kiribati
Nauru
Uganda
Kenya
Rwanda
Burundi
United Republic of Tanzania
Seychelles
Indonesia
Papua New Guinea
Solomon Is.
Timor-Leste
Christmas (Australia)
Cocos (Keeling) Islands (Australia)
Comoros
Mozambique
Mauritius
Réunion (Fr.)
Vanuatu
New Caledonia (Fr.)
Australia
Swaziland
New Zealand
wfp vam
vulnerability analysis and mapping

1.2 Who are the most vulnerable?

"Children are the living messages we send to a time we will not see."

John W. Whitehead, 1983

Hunger and poor health are not shared equally among all people; the burden falls largely on the marginalized poor, with further disparity by gender, age and ethnicity. The life cycle impacts of hunger and poor health can be profound when compromised health spans generations. Undernourished women are most likely to have low birthweight infants, who in turn are likely to have less chance of survival and experience poor health through their lives (McCormick, 1985; Barker, 1998). The cumulative effects of hunger and poor health during childhood are reflected in later life, marked by the adult prevalence of micronutrient deficiencies, chronic diseases such as diabetes and obesity, hypertension, and other conditions. The implications of hunger across the life cycle thus extend from conception to old age, with human capital formation and well-being undermined at each stage.

Women and mothers

Women often have different and unequal access to and use of basic health resources, including primary health services for the prevention and treatment of diseases. Healthcare for women and their nutritional needs are too often not given sufficient priority, and women's lack of participation in decision-making at home and in the community further precludes meeting women's nutrition and health requirements. Women often lack adequate nutrition for their own growth, development and survival.

Women of child-bearing age have greater nutrition and health needs than other age and gender groups, primarily because of the high physiological demands of pregnancy and childbirth. Paradoxically, cultural and social biases often lead to the presumption that women have lower nutritional needs. Women's physical activities and caloric expenditures consequently tend to be underestimated, hence the observation "women eat less and last". The micronutrient needs of women, particularly women of child-bearing age, are also often neglected.

Inadequate nutrition among pregnant women impacts children's growth and development throughout the life cycle. The stature of mothers, their nutritional status before and during pregnancy and their weight gain during pregnancy have all been shown to have significant impacts on the growth and development of the foetus. Short stature among women, a symptom of past hunger, has been associated with a higher risk of pre-term and difficult births (Siega-Riz et al., 1994; Nestel, 2000). A mother's health also affects foetal development, especially if she has experienced diarrhoeal diseases, malaria, parasitic infections or respiratory infections during pregnancy.

The continuing high rate of maternal deaths in developing countries has been attributed to the "three delays": delays in recognizing that complications are serious enough to require help; transport constraints in getting to a treatment centre equipped for obstetric emergencies; and delays in receiving treatment because of lack of available and trained healthcare personnel, life-saving drugs and/or equipment (Global Health Council, 2007).

Eliminating hunger in women and mothers is a key entry point to break the inter-generational cycle of hunger.

Infants and young children

Children between the ages of 6 months and 3 years are particularly vulnerable to hunger. Vulnerability increases as breast milk consumption is reduced and complementary foods are introduced (World Bank, 2006). The quantity and quality of food at this critical stage is vital to the growth of the child.

Care practices are an essential counterpart to the food a child receives. Childcare practices are provided mostly but not exclusively by women; they range from food preparation and food storage, breastfeeding and the feeding of very young children to hygiene practices, health-seeking behaviour and psychosocial stimulation of children (Engle, 1999). Breastfeeding is the critical practice of providing food (breast milk), health (transfer of active immunity) and care

Hunger and poor health from birth: typical factors

- Pregnant and nursing women eat too few calories and too little protein, have untreated infections that lead to low birthweight, or do not get enough rest.
- Mothers have too little time to take care of their young children or themselves during pregnancy.
- Mothers of newborns discard colostrum (the first milk, which strengthens the child's immune system).
- Mothers feed children under age 6 months with foods other than breast milk, even though exclusive breastfeeding is the best source of nutrients and the best protection against many infections and chronic diseases.
- Caregivers start introducing complementary solid foods too late.
- Caregivers feed children under 2 with too little food, or foods that are not energy dense.
- The needs of women and young children are not met, even though food is available, because of inappropriate household food allocation, and their diets often do not contain adequate micronutrients or protein.
- Caregivers do not know how to feed children during and after diarrhoea or fever.
- Poor hygiene practices contaminate food with bacteria or parasites.

Source: World Bank, 2006

(stimulation and security) to the breastfed child (Lindstrand et al., 2006).

Access to resources is also critical. When women control household income, for example, they may direct larger proportions of food to children. The amount of time mothers can devote to care practices is also important. Whether a mother's work affects a child's nutrition adversely or positively depends on several factors, including the availability of alternative caregivers.

Infants and young children are a priority intervention point for breaking the cycle of hunger.

School-age children and adolescents

In children under 5 in developing countries, short stature or stunting is highly prevalent and affects one child in every three, an estimated 178 million children (WHO, 2007e). The opportunity for catch-up is minimal during childhood and adolescence, especially if children remain in the same environment. As detailed in the *World Hunger Series 2006*, poor health and nutrition during early childhood also impacts cognitive development and school performance, with negative consequences continuing into adulthood.

Adolescence is marked by intense physical, psychosocial and cognitive development. Adolescents are particularly vulnerable to hunger because of the combined effects of increased physical activity, poor eating habits and/or lack of access to nutritious foods, and unequal access to safe and improved livelihood opportunities. However, adolescents generally experience lower rates of morbidity and less susceptibility to disease and face fewer life-threatening conditions than children under 5.

Ages 10–19

Adolescents are people aged 10–19 years; they comprise 20 percent of the world's population (WHO, UNFPA and UNICEF, 1995).

Adolescent health is closely intertwined with hunger, especially in situations where:

- The nutrition transition is linked to increasing rates of obesity and chronic diseases.
- Conflict and war prevail and adolescent boys are targeted to serve as soldiers or join rebel groups and girls are raped or forced into long-term sexual slavery.
- The AIDS pandemic has been an uncurbed force; the plight of AIDS orphans, a tragic manifestation of the worldwide pandemic, deserves special attention. These children are extremely vulnerable to a range of nutritional, health, psychosocial and economic risks.
- Rapid urbanization in the developing world is resulting in growing slums and, for young people, high rates of unemployment, suicide and crime (Blum, 1991).

- Human trafficking, most often for the sex trade and domestic service, is prevalent.

Adolescence is a crucial intervention point from a hunger and health perspective – a period during which significant progress can be made in improving maternal and child health and in setting the stage for a healthy adult life.

Underweight and adolescent girls

Underweight adolescent girls may not have completed growing before their first pregnancy. Growing adolescents are likely to give birth to smaller babies than mature women of the same nutritional status because of poorer placental function and competition for nutrients between the growing adolescent and the growing foetus (Gillespie, 2001).

The elderly

The 20th century has seen an unprecedented global transition from high birth and death rates to low fertility and mortality. Although people live longer, a reduced role in the labour market constrains the elderly. Economic constraints can result in their accessing inadequate and poor quality food. Ageing people also face social stigmatization that affects their access to timely healthcare, which can be detrimental because they tend to have greater healthcare needs and weaker immune systems and thus are more susceptible to new infections.

Research carried out by the London School of Hygiene and Tropical Medicine and HelpAge International found a high prevalence of undernutrition in older adults; it was highest among the very old in India, Malawi and Rwanda (Ismail and Manandhar, 1999). Also, little is known about the micronutrient status of older people in developing countries. The type of nutrition problem affecting the elderly depends on the relative prosperity of the country, equitable distribution of economic resources and the extent to which public resources are used for health and social welfare programmes, especially those involving nutrition (Bermudez and Dwyer, 1999).

The changing role of grandparents may affect their health and nutritional status. In poor households, grandparents often tend to reallocate resources away from personal needs to meet the basic requirements of their grandchildren. Thus the risk of poor health and inadequate food among older family members becomes higher.

Elderly people have a reduced capacity to care for their children and extended family. However, caring for those affected by HIV/AIDS often falls to the elderly who have waning capacity to take on additional responsibilities. Older women in particular play a critical role in caring for AIDS orphans (UNICEF, 2004).

The elderly must not be forgotten when designing hunger and health interventions.

Refugees and displaced populations

There were about 3,300 natural disasters between 2000 and 2007. In 2006 alone, more than 143 million people were affected by 426 natural disasters (CRED, 2007a). Conflict also takes its toll: 20.8 million people, predominantly in Africa and Asia, are still displaced. Of the ten countries with the world's highest rates of under 5 mortality, seven are affected by conflict or its aftermath (UNICEF, 2006b). The actual and potential impact of crises is enormous; the human suffering is immense, and gross national income (GNI) declines.

Conflicts and natural disasters disrupt food supplies and health services and cause water and sanitation systems to break down. They also create high levels of stress among families and communities. Refugees and displaced populations are among the most vulnerable because they leave behind their land, livelihoods and property and do not have the means to meet their own food or health needs.

Even internally displaced people (IDPs) and refugees who seek refuge in camps are not guaranteed fully adequate services. In many cases their nutritional and health status declines as a result of inadequate shelter, overcrowding, shortage of medicines and insufficient water and sanitation services.

Figure 6 – Natural disasters worldwide

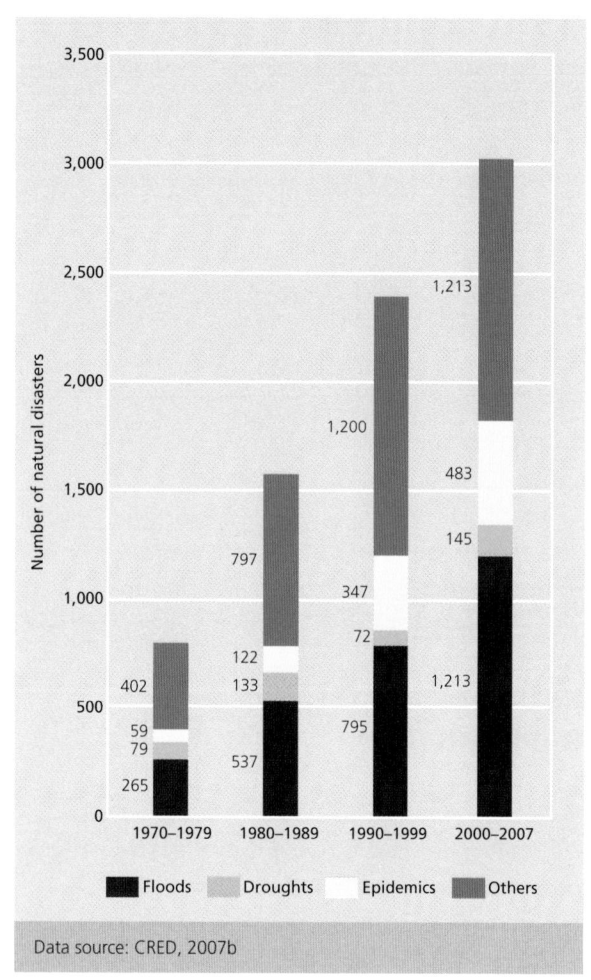

Data source: CRED, 2007b

Deaths are rarely directly related to food shortages, rather more commonly to the interaction between infectious disease and pre-existing undernutrition. Specifically, lack of food does contribute to mortality because inadequate nutrients make people more susceptible to recurring and more severe episodes of illness. Bringing people together, especially rural people, can expose them to a range of infectious diseases for which they have not acquired immunity. Avoiding large-scale population movements is fundamental to avoiding hunger, disease, loss of livelihoods and human suffering. The major causes of morbidity and mortality among refugees are measles, diarrhoeal diseases, acute respiratory infections, malaria and undernutrition (Mason, 2002). These consistently account for 60–95 percent of all reported causes of death among the displaced (Waldman, 2005).

Close attention to hunger as part of early-warning, relief and rehabilitation efforts can help to improve the outcomes of humanitarian assistance. Knowledge of the initial nutritional condition of the affected population is critical to averting hunger-related morbidity and mortality, and choices that improve nutrition and health service coverage can reduce vulnerability to disease and death in times of crisis.

Nutritional surveillance as part of crisis prevention and disaster mitigation can help to pinpoint geographic areas that are more vulnerable to crisis and help to target assistance efficiently once a population is affected.

Indigenous people

Although nutritional status and life expectancy have improved in developing countries, indigenous groups have not kept pace with global progress. Higher infant and child mortality trends reveal inequity in access to adequate food and healthcare for many indigenous people.

For example, in Guatemala the national prevalence of chronic undernutrition is 46.4 percent, but the rates vary dramatically across the country, and more severe pockets of hunger are easy to locate. The percentage of stunting (height-for-age) in children under 5 reaches 94 percent in some remote departments, yet in metropolitan areas the figure drops to 35.7 percent, which is still high. Areas where chronic undernutrition is highest are often the most isolated and populated by marginalized or indigenous groups.

Guatemala has one of the widest income gaps in the world: the top 10 percent of the population receives 50 percent of national income, while the bottom 50 percent receives little more than 10 percent (IDB, 1999). Guatemala is not on target to achieve the MDG hunger goal of halving the proportion of undernourished people by 2015; it is also one of only four countries in the Latin America and the Caribbean Region (LAC) expected to miss the target of halving the number of underweight children under 5. (Data were available for 22 of the 33 countries in the region.)

Other examples where indigenous groups are falling behind include (PAHO, 2007):

- El Salvador: 91.6 percent of the indigenous population obtain water from wells, rivers or both. Only 33 percent of indigenous people have the benefit of electricity; 64 percent use oil lamps or candles for lighting.
- Mexico: Infant mortality among indigenous people was 59 per 1,000 live births in 1997, twice the national infant mortality rate.
- Peruvian Amazon region: 32 percent of the population in indigenous communities are illiterate, compared with 7.3 percent for the total population.
- The United States of America: Indigenous people are much more likely than the general population to die from diabetes mellitus related to obesity, and liver disease resulting from alcohol abuse.
- Australia: Aboriginal and Torres Strait Islander peoples have a much shorter life expectancy than the general Australian population. Indigenous Australians born between 1996 and 2001 are expected to live nearly 20 years less than the rest of the population (Australian Institute of Health and Welfare, 2007).

Chronic undernutrition/stunting and indigenous groups in Guatemala

Department	% Indigenous population	% Stunting in children under 5
Huehuetenango	69.3	72.0
Quiche	64.5	80.0
Totonicapan	72.9	94.5
Quetzaltenango	60.0	68.2

Data source: WFP, 2005

New obligations are arising for governments in terms of recognizing, promoting and guaranteeing the individual and collective rights of indigenous people in line with international standards.

To close gaps and achieve the MDGs with equity requires an inter-cultural perspective that increases coverage through healthcare systems and respects the different food systems of the world's people.

Comprehensive Food Security and Nutrition Survey in Liberia

Battered by civil war from 1989 to 2003, Liberia is now on a long road to recovery. In order to plan for that recovery it is essential to understand people's hunger and health status, the causes of food insecurity and undernutrition, livelihood patterns and agricultural constraints.

To identify those most vulnerable to hunger, a joint Comprehensive Food Security and Nutrition Survey was undertaken between February and June 2006 led by the Government of Liberia, in collaboration with United Nations agencies and non-governmental organizations (NGOs). The survey found that 39 percent of children under 5 are stunted or too short for their age; 6.9 percent of children under 5 are wasted or too thin for their height, and 27 percent of the children under 5 are underweight. Also, 50 percent are highly vulnerable to hunger or vulnerable to become food insecure.

In 9 of the 15 counties surveyed, the rate of stunting is higher than 40 percent – rates above 40 percent indicate a critical situation – whereas the remaining counties show serious levels of chronic malnutrition (30–40 percent). In the central and south-eastern counties the rate of wasting among under 5 year olds is over 10 percent, indicating the need for immediate action. In all counties, 12 percent of children aged 12–24 months were affected by acute malnutrition and a high prevalence of associated illnesses, including diarrhoea, malaria and acute respiratory infections. Poor infant and child feeding practices were common.

Families returning to their homes are particularly vulnerable to hunger as they begin to restore livelihoods destroyed by war. Households with particular difficulties in accessing sufficient food, by either purchase or their own production, include those:

- headed by women and the elderly;
- with chronically ill or disabled household members;
- of larger size and those that are overcrowded;
- with vulnerable livelihoods (including those that are palm oil producers and contract labourers); and
- without access to agricultural land.

Communities have limited access to health services. Overall, 90 percent of communities reported not having a local health facility – on average communities without a local health facility reported walking for nearly three hours to the nearest facility.

The health facilities are mainly funded and managed by NGOs although owned by the Government; 18 percent are run by private institutions or individuals and 14 percent are run or funded by the Government. The findings are not surprising considering that the previous conflicts had destroyed much of the rural infrastructure and greatly reduced the Government's ability to provide adequate healthcare.

Source: excerpt from WFP, 2006a

Map 2 – Hunger and natural disasters

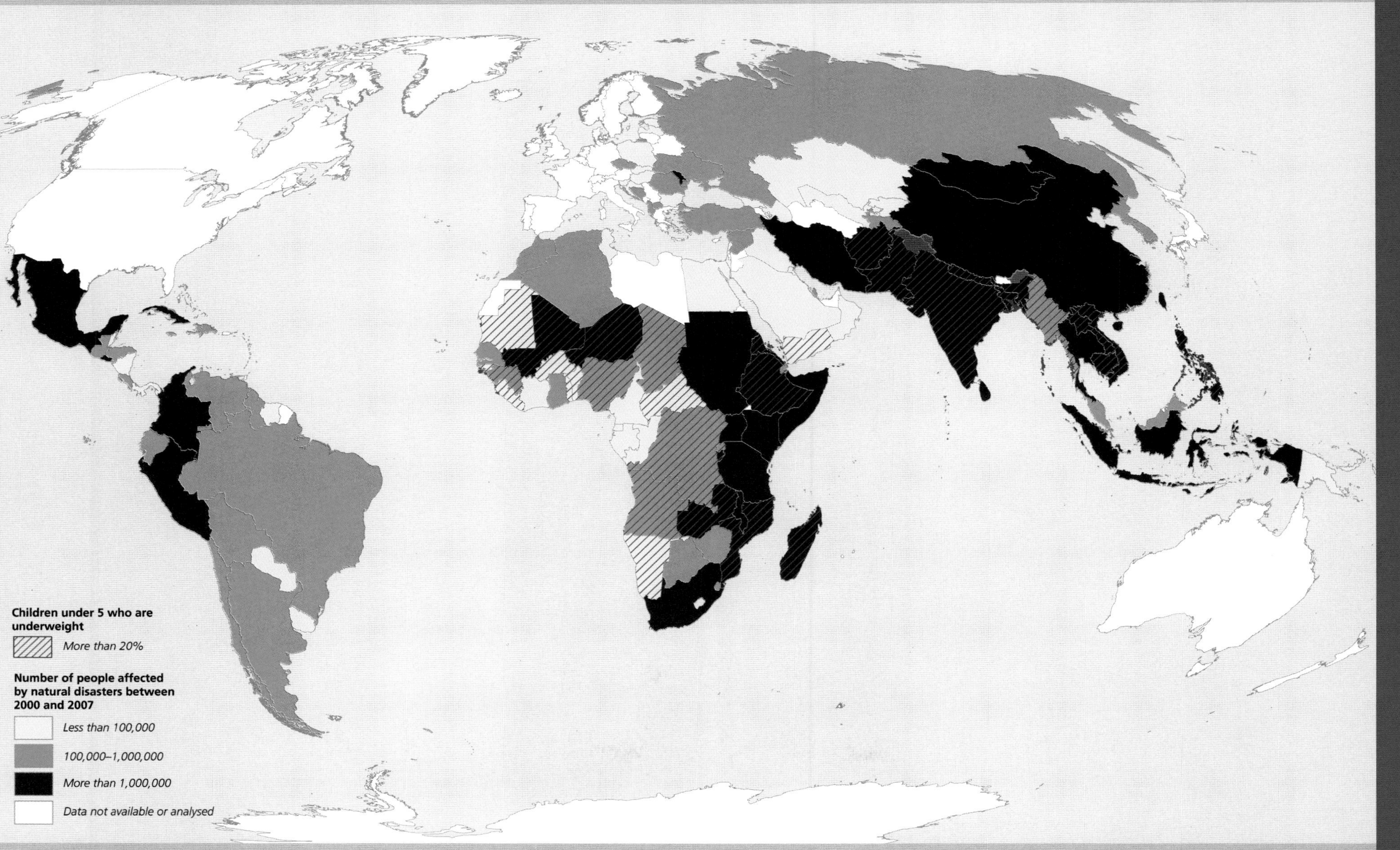

The boundaries and the designations used on this map do not imply any official endorsement or acceptance by the United Nations. Map produced by WFP VAM.

Data sources: Centre for Research on Epidemiology of Disasters, 2007; WHO, 2007

wfp vam
vulnerability analysis and mapping

Intermezzo 2: Women and ending hunger – the inextricable link

The subjugation, marginalization and disempowerment of women is particularly severe in the countries where hunger persists. This link is powerfully illustrated in South Asia.

Currently, India has millions of tons of surplus food in storage – yet India is home to the largest number of hungry people in the world. In fact, India and Bangladesh account for nearly one third of the remaining hunger in the world. The South Asia region has the world's highest rate of child malnutrition: nearly one third of babies born in South Asia are underweight and malnourished; this compares with 14 percent in sub-Saharan Africa.

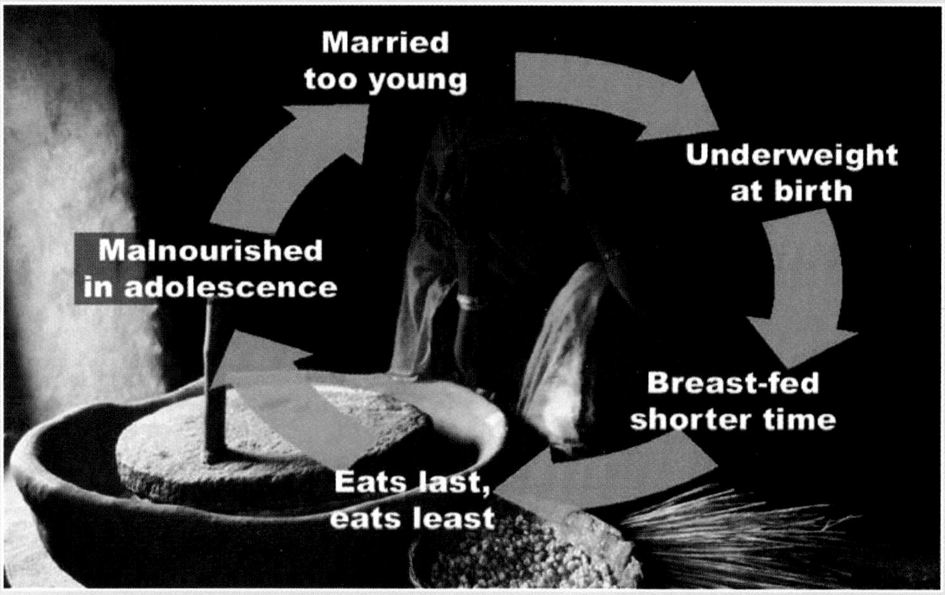

To understand this anomaly, in 1996 the United Nations Children's Fund (UNICEF) published a landmark study by India's leading nutritionist entitled *The Asian Enigma*. The study concludes that "… the exceptionally high rates of malnutrition in South Asia are rooted deep in the soil of inequality between men and women".

The study found that girls and women in South Asia are less well cared for than girls and women in sub-Saharan Africa. In South Asia, women and girls eat last and least: they eat only the food that is left over after the men and boys in the household have eaten. Often, men and boys consume twice as many calories, even though women and girls do much of the heavy work.

It is widely recognized that the health and nutritional status of a pregnant woman dramatically affects the health of her baby. New scientific data make it clear that it is not just about her health when she is pregnant, or even throughout her life: it goes back to when she herself was in the womb. It is now evident that an insidious "cycle of malnutrition" exists in the regions where hunger persists. This is particularly true for South Asia.

This cycle of malnutrition begins when a baby girl is born underweight and malnourished. She is breastfed less and fed less nutritious food than her brother. She is often denied healthcare and education. She is forced to work even as a child. Her work burden increases significantly as she gets older, even when she is pregnant. She is married and pregnant when she is young, often just a teenager. She is underweight and malnourished when she gives birth to her children, who are born underweight and malnourished. And so the cycle continues.

This deprivation of women and girls has a profound effect on society. When children are born malnourished and underweight, they are at severe risk in all areas of personal development, health and mental capacity. They are physically weak and lack resistance to diseases such as tuberculosis and malaria. They face a lifetime of disabilities, a reduced capacity for learning and diminished productivity.

New research shows that maternal deprivation before and during pregnancy also makes a woman's body susceptible to diseases we associate with affluence, for example hypertension, cardiovascular diseases and type II diabetes.

Current figures show the magnitude of the problem: for example, cardiovascular disease is the leading cause of mortality in India; and in the next 20 years, India will have the largest number of diabetic patients in the world – 79,440,000, nearly 22 percent of the world's diabetics.

The research into the subjugation, marginalization and disempowerment of women throughout their lives underscores the undeniable link between the neglect and discrimination of women and girls and adverse effects on the health and survival of all.

It is clear that traditional responses to child malnutrition, such as providing nutritional supplements for pregnant women, are inadequate. To interrupt the cycle of malnutrition, a woman's health and nutrition must improve throughout her entire life. This means transforming the way women and girls are treated in the family and in society as a whole.

The link between gender and hunger goes beyond nutrition. Women bear almost all the responsibility for work related to family health, education, nutrition and – increasingly – for family income. Yet women are systematically denied the education, resources and voice in decision-making that they need to fulfil these responsibilities. Conversely, when women achieve social, education and political progress, the welfare of the entire family improves. A study of progress in nutrition over 25 years showed that the single most important factor was the education of women.

Gender has long been recognized as an important factor in hunger. But it is now clear that it is the fundamental cause of most of the remaining hunger in the world.

Contributed to the World Hunger Series *by Joan Holmes, President of the Hunger Project. Photo by Andrea Booher, United Nations Development Programme (UNDP).*

1.3 Tracking the hunger and health MDGs

"Achieving MDGs at the national level is not the same as achieving MDGs for all: the global development community has largely focused on the national level in the context of global reporting. Within countries there is an urgent need to address progress among particular groups and areas."

Davidson R. Gwatkin, 2005

The MDGs are quantified targets for addressing poverty in its several forms: hunger, income, illness and lack of shelter, and promoting gender equality, education and environmental sustainability. They also encompass the basic human rights of food, health, education, shelter and security. The first MDG relates explicitly to reducing hunger. Achieving reductions in hunger as embodied in MDG 1 will directly benefit all of the other MDGs, in particular those related to health. The table below illustrates the critical relationship between eliminating hunger and achieving the MDGs.

Progress on meeting the MDGs

Current progress in achieving the MDGs is uneven, and insufficient. Throughout the world there are countries lagging behind, and in each country there are poor, hungry people – women and girls, refugees, IDPs and other vulnerable groups living in forgotten areas – who do not have sufficient access to adequate food, healthcare, education or water and sanitation.

The MDGs and hunger

MDG	Relationship with health and hunger
1. Eradicate extreme poverty and hunger	Hunger erodes human capital, often irreversibly, throughout the life cycle. It impairs health and livelihood opportunities. Hunger linked to disease is a major cause of mortality and morbidity in conflicts and natural disasters.
2. Achieve universal primary education	Hunger affects the chances that a child will go to school, stay there and perform well. The long-term consequences and inter-generational impact are widely recognized.
3. Promote gender equality and empower women	Hunger and undernutrition negatively affect women's health and their ability to care for their children and family. It limits their livelihood opportunities. Empowerment of women is essential to achieve the MDGs.
4. Reduce child mortality	Undernutrition caused by the dual effects of hunger and infectious disease is the leading cause of child mortality.
5. Improve maternal health	Hunger is associated with most major risk factors for maternal morbidity and mortality. Stunting and micronutrient deficiencies put mothers at higher risk for complications during pregnancy.
6. Combat HIV/AIDS, malaria and other diseases	Hunger increases the risk of HIV transmission, compromises anti-retroviral therapy and hastens the onset of AIDS. It increases the chances of tuberculosis (TB) infection and reduces malaria survival rates.
7. Ensure environmental sustainability	Reducing hunger and improving health are linked to improving access to clean drinking water and sanitation. Hungry people are less able to care for the environment in a sustainable way. They may be forced to exploit their natural resource base in order to meet immediate food needs.
8. Develop a global partnership for development	Hunger needs to be systematically addressed in the context of the other MDGs, in international development and humanitarian programmes and with respect to international trade.

Sources: based on WFP, 2006b; FAO, 2005; World Bank, 2006

1.3 Tracking the hunger and health MDGs

"Between 5 and 6 million children die each year from infectious diseases that would not have killed them if they had been properly nourished. The weekly child death toll from hunger and undernutrition far exceeds those caused by even the most dramatic natural disasters. ... A global political consensus has emerged on the need to tackle hunger as being fundamental for achieving the MDGs."

Ending Child Hunger and Undernutrition Initiative, 2006

As monitoring progress towards meeting the MDGs is carried out at the national level, it is not always easy to track sub-national advances or progress for certain vulnerable groups. The available data are not updated annually even for national monitoring; consequently sub-national assessments must rely on available country-specific surveys. The Guatemala example discussed previously vividly demonstrates that national figures tend to mask the disparities that exist within countries, as well as trends and changes – both positive and negative – in and among countries. Further sources do not allow for a breakdown of national data to examine who is benefiting from social programmes and who is falling behind.

The "Hunger Hotspots" approach of the United Nations Millennium Project's Hunger Task Force is an innovative method for identifying sub-national areas where more than 20 percent of pre-school children are underweight. This approach supports the premise that hunger must be measured and analysed at the sub-national level – state, province or district – where hunger may be deeply rooted (United Nations Millennium Project – Hunger Task Force, 2005).

Another useful tool for identifying populations vulnerable to hunger at the sub-national level is WFP's vulnerability analysis and mapping (VAM). While beneficial for targeting populations in need and identifying appropriate interventions, VAM also provides a baseline for tracking MDG indicators. VAM helps to enhance hunger analysis at the sub-national level and identifies those most in need of assistance.

Hunger progress: on track or not?

At the midway point to achieving the MDGs by 2015, there have been several assessments to see if progress is on track. The results vary according to data sources and the background of the research undertaken, including the grouping of countries into regions (FAO, 2006a; UNICEF, 2006a; WHO, 2005; World Bank, 2007).

Although there are many trends that show improvements in reducing hunger in some countries and for selected groups, these assessments indicate that global attainment of MDG 1 target 2 is not on track. More distressing still is the fact that in some parts of the world past progress is being eroded and sustainable solutions are still far off for the hungriest people. South Asia and sub-Saharan Africa are most seriously behind.

This section presents an updated analysis of progress towards meeting MDG 1 target 2 – sometimes referred to as the hunger target – and selected health targets closely aligned to hunger. Using World Health Organization (WHO) 2007 data and Food and Agriculture Organization of the United Nations (FAO) 2006 data, the analysis paid particular attention to LIFDCs because of their high vulnerability to hunger and the special challenges these countries face in meeting MDG 1.[4]

MDG 1: Eradicate extreme hunger and poverty

Target 2. Halve, between 1990 and 2015, the proportion of people who suffer from hunger.

The target is measured by two indicators:

- prevalence of underweight children under 5;
- proportion of population below minimum level of dietary energy consumption (undernourishment).

Findings are consistent with other assessments and confirm that, globally, trends are mixed and that advances are still not sufficient to reach the 2015 target (see Table 10 on page 173 in the Resource Compendium for country-level data). Specifically, LIFDCs tend to lag behind other developing countries in eliminating underweight and undernourishment, the two main indicators used to track progress.

The prevalence of underweight in children under 5 follows a similar pattern in Africa, with the largest proportion found in sub-Saharan Africa. In South Asia, however, underweight is much worse than undernourishment. While these two measures show overwhelmingly that the poorest countries face the greatest challenges in overcoming hunger, hunger still persists in the world's richest countries.

Regional reductions in underweight children under 5

The prevalence of underweight children under 5 declined from 33 percent to 27 percent in developing countries between 1990 and 2005 (United Nations, 2007). However, progress is not equal: the prevalence of underweight children in LIFDCs is still about 30 percent, and 13 of the 54 LIFDCs for which data were available are regressing on the underweight indicator. The countries making the greatest progress are Bhutan and China; Albania and Yemen have the slowest progress. The total number of underweight children under 5 in LIFDCs is 121 million, compared with 143 million in all developing countries.

There are significant variations within regions, particularly Asia, where China's rapid progress to reach the indicator offsets the slow progress in India and elsewhere. In South East Asia, the prevalence of underweight children under 5 is almost 45 percent among LIFDCs, the highest in the world. The absolute number of children affected also remains unacceptably high.

In African LIFDCs, the prevalence of children under 5 who experience severe or moderate underweight is 30 percent compared to the target of 10 percent or lower. About 64 percent of African LIFDCs reporting on the underweight indicator are not on target. In Latin America the prevalence in LIFDCs has fallen to 15 percent.

Regional reductions in undernourishment

It is estimated that there are 854 million people whose food intake is not sufficient to meet their basic energy needs – 820 million in developing countries, 25 million

Figure 7 – Prevalence of child underweight in LIFDCs by region

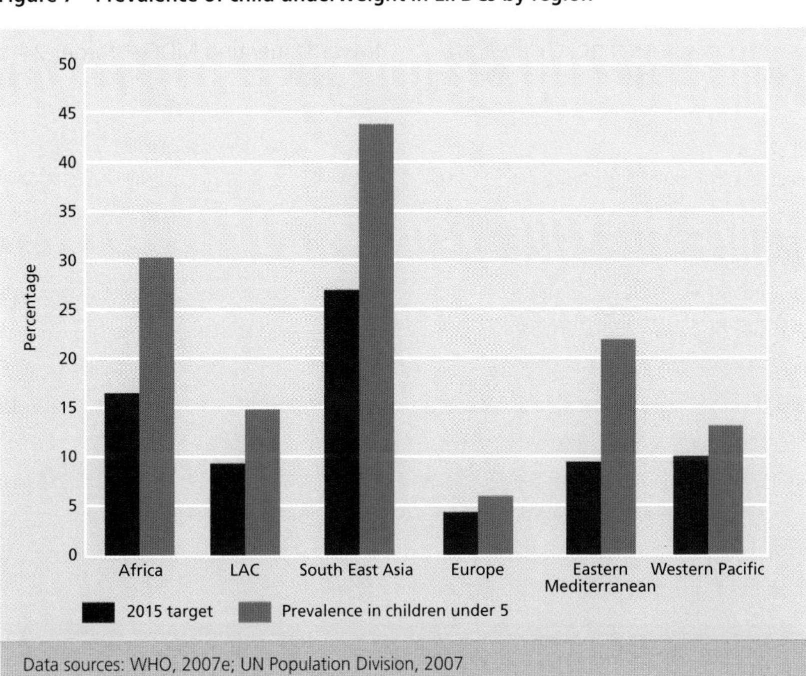

Data sources: WHO, 2007e; UN Population Division, 2007

Figure 8 – Prevalence of undernourishment in LIFDCs by region

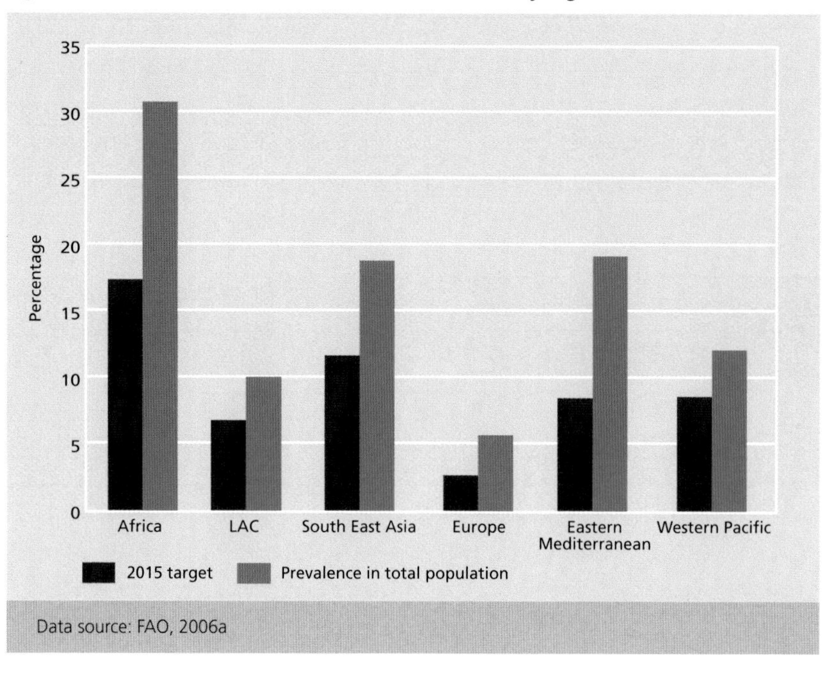

Data source: FAO, 2006a

in countries in transition and 9 million in industrialized countries (FAO, 2006a). More than half of all undernourished adults and children are in Asia and the Pacific, whereas the highest proportion is found in sub-Saharan Africa. Long-term trends show that the proportion of undernourished people has fallen markedly: from 37 percent of the total world population in 1969–1971, to 20 percent in 1990, and to 17 percent in 2001–2003 (FAO, 2006b). However, 86 percent of the global undernourished population live in LIFDCs, which translates into 726 million undernourished people in the world's most food-insecure countries.

Progress towards the MDG target is also uneven. There has been progress in all regions and in most countries, but a number of countries have regressed and some have suffered serious setbacks. LIFDCs that have suffered setbacks include: Burundi, the Democratic Republic of the Congo (DRC), The Gambia, Tanzania and Yemen. Fourteen LIFDCs show some level of increase in the proportion of undernourished people.

In sub-Saharan Africa the gains have been offset by population growth, causing a large increase in the absolute number of undernourished individuals and large sub-regional variations. "Regional aggregate trends however, conceal significant sub-regional differences. Within sub-Saharan Africa, Central Africa has made the least progress in reducing the prevalence of undernourishment" (FAO, 2006b).

According to FAO, the major gains in combating undernourishment have occurred in Asia and the Pacific, where there has been a 66 percent reduction.

Some of the factors impeding progress in terms of the two hunger indicators include high HIV/AIDS infection rates, political instability, armed conflict and an increase in natural disasters over the past few years. Conflict and the forced movement of refugees or IDPs significantly increase Africa's vulnerability to hunger.

Conversely, some of the country characteristics stimulating progress towards the hunger goal include growth in the agricultural sector, openness to equitable trade and political stability.

Hunger progress overall

Using both indicators to track the hunger target, out of the 70 developing countries analysed of which 47 are LIFDCs:

Figure 9 – Hunger indicators by region

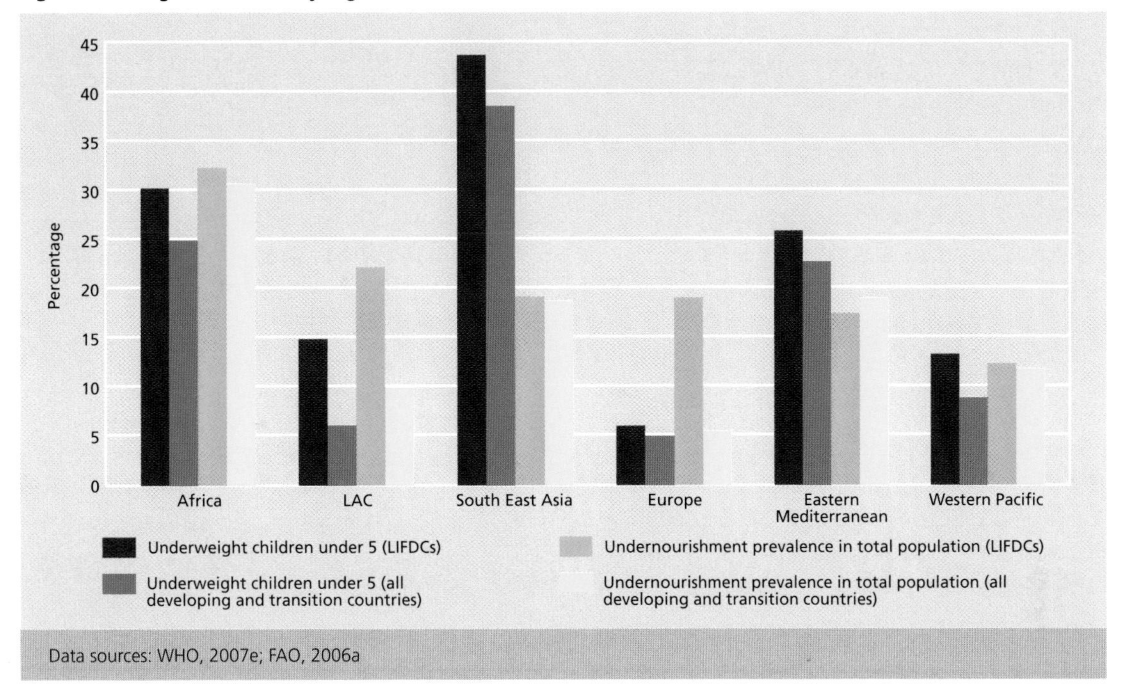

Data sources: WHO, 2007e; FAO, 2006a

- 19 developing countries, of which 10 are LIFDCs, are on track to meet both hunger indicators;
- 2 LIFDCs are regressing on both indicators – Burundi and Yemen;
- 7 countries are falling back on the underweight target, of which all are LIFDCs; and
- 11 countries are regressing on the undernourishment target; 7 of these are LIFDCs.

Progress on meeting the health MDGs

The latest WHO data show that progress towards meeting health-related MDGs, as for hunger, is also uneven. Economically well-off countries are improving their health at a faster rate than those that are worse-off.

MDG 4: Reduce child mortality. Progress lags behind the other MDGs. In 2005, only 84 of 163 countries were on track to achieve the MDG 4 child mortality target and were making sufficient progress to achieve a 67 percent reduction by 2015. Again we see wide regional variation:

- 45 percent of all child deaths occur in sub-Saharan African countries. Every year, 4.8 million children in sub-Saharan Africa die before the age of 5. It is the only region in the world where the number of child deaths is rising (UNDP, 2005).
- The number of African children at risk of dying is 35 percent higher than it was 10 years ago (Gordon et al., 2004).
- Among the worst performing LIFDCs are Cambodia, Côte d'Ivore, Iraq, Swaziland and Zimbabwe.
- In contrast, child survival has improved in Latin America and the Caribbean, South East Asia and northern Africa, where child mortality rates have declined by more than 3 percent. Albania, Egypt, Indonesia, Syria and Timor-Leste are making rapid progress and are on track to achieve MDG 4.

An estimated 63 percent of child deaths could be averted with basic healthcare and treatment (World Bank, 2007).

MDG 5: Improve maternal health. Trends indicate general improvements throughout all regions; however, 99 percent of maternal deaths – 500,000 per year – occur in developing countries (World Bank, 2007). Data collection is uneven for maternal mortality, so the proxy "skilled attendance at delivery" is sometimes used. According to the World Bank, gaps in access to skilled attendance are higher than for any other health or education service. Few countries have sufficient data to show well-documented progress.

Of the 136 countries analysed, however, 91 show progress, but it is highly concentrated in richer countries and households. Among LIFDCs, 24 countries are regressing and 25 are progressing but are not sufficiently on track to achieve MDG 5. Azerbaijan, Malawi, Mongolia, Tanzania and Zimbabwe are among the poorest performers. Bhutan, Indonesia, Morocco, Papua New Guinea and Vanuatu are some of the most rapidly progressing countries and are on track to achieve MDG 5.

Only 46 percent of births are assisted by professional health workers in sub-Saharan Africa, where the number of maternal deaths accounts for more then 50 percent of global cases (UNICEF, 2006a).

More attention needs to be given to the food and healthcare needs of women: before, during and immediately after pregnancy.

MDG 6: Combat HIV/AIDS, malaria and other diseases. Despite international and national efforts, the HIV burden rose from 37.1 million people in 2004 to 39.5 million in 2006, a move away from the MDG 6 target. The HIV/AIDS situation is devastating: in 2006 alone 3 million people died, more than ever before and more than from any other infectious disease. Another 4.3 million people became infected in 2006 (UNAIDS, 2006; WHO, 2007b).

The HIV/AIDS epidemic is highly concentrated in sub-Saharan African countries; 64 percent of all HIV-positive people and 90 percent of children under 15 with HIV live in sub-Saharan Africa. While the spread of the disease has slowed in sub-Saharan Africa, HIV/AIDS is a rapidly growing epidemic in Eastern Europe and Central Asia (World Bank, 2007). Prevention and treatment programmes are still limited: only 8 percent of those who need anti-retroviral therapy (ART) in the developing world receive it; only 4 percent in sub-Saharan Africa will have access to treatment, and only 8 percent of pregnant women are offered services for preventing transmission to their infants.

The United Nations Millennium Declaration sets three main health-related goals:

MDG 4: Reduce child mortality.
MDG 5: Improve maternal health.
MDG 6: Combat HIV/AIDS, malaria and other diseases.

This report reviews the following set of indicators for health that directly relate to hunger and poor health:

- under 5 mortality rate (MDG 4);
- infant mortality rate (MDG 4);
- maternal mortality ratio (MDG 5);
- proportion of births attended by skilled health personnel (MDG 5);
- HIV prevalence among pregnant women aged 15–24 (MDG 6);
- prevalence and death rates associated with malaria (MDG 6); and
- prevalence and death rates associated with TB (MDG 6).

Tuberculosis (TB) caused 1.6 million deaths in 2005, including 195,000 patients infected with HIV, and rates continue to rise. In 2005 there were nearly 8.8 million new TB cases, 7.4 million in Asia and sub-Saharan Africa (WHO, 2007d). Each person with active TB will infect 10–20 people every year on average (WHO, 1997). The total number of patients diagnosed and treated in 2006 was similar to 2005 (WHO, 2007e). Approximately a third of the 40 million people living with HIV/AIDS are also infected with TB. People with HIV are up to 50 times more likely to develop active TB in a given year (WHO, 2007c).

Although TB incidence, prevalence and death rates now appear to be in decline, rates are not yet falling fast enough to achieve the 2015 targets; in Africa rates increased dramatically between 1990 and 2005.

Approximately 40 percent of the world's population, mostly those living in the world's poorest countries, are at risk of malaria. Every year, more than 500 million people become severely ill and more than 1 million people die from malaria – mostly infants, young children and pregnant women in sub-Saharan Africa (UNICEF, 2007). Asia, Latin America, the Middle East and parts of Europe are also affected. Large and devastating epidemics can occur in areas where people have had little contact with the malaria parasite and therefore have little or no immunity. These epidemics can be triggered by weather conditions and further aggravated by complex emergencies or natural disasters.

While access to effective treatment and prevention methods have been increasing, current trends in malaria morbidity and mortality are not yet showing sufficient progress towards achieving the 2015 targets.

Data collection needs to be intensified to improve the tracking of progress on all the MDG indicators, especially MDG 5 and 6.

Figure 10 – Progress on underweight: MDG 1 for developing LIFDCs, 1990–2006[5]

Figure 11 – Progress on undernourishment: MDG 1 for developing LIFDCs, 1990–2003[5]

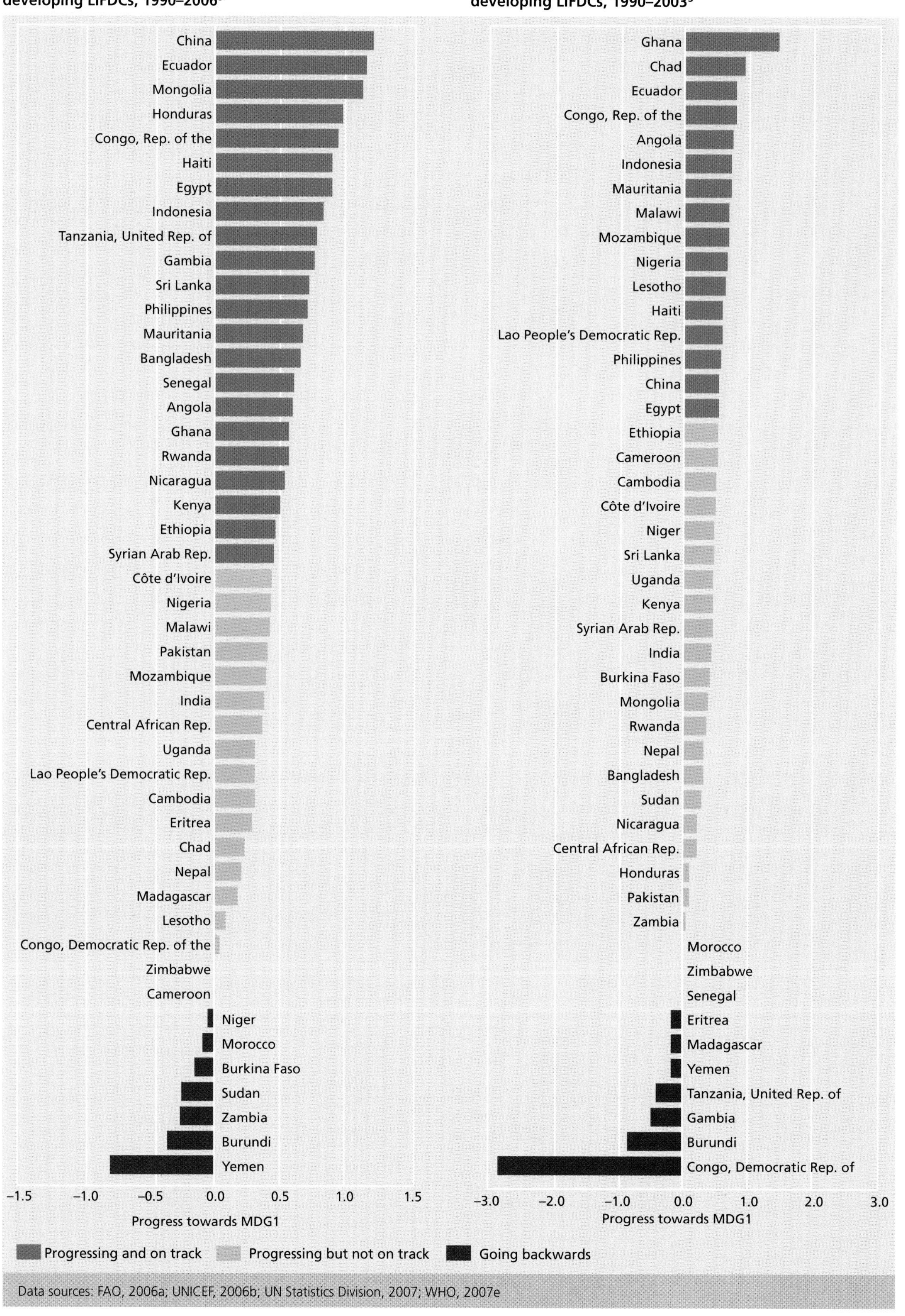

Data sources: FAO, 2006a; UNICEF, 2006b; UN Statistics Division, 2007; WHO, 2007e

Figure 12 – Progress on child mortality: MDG 4 for developing and transition LIFDCs, 1990–2005[5]

Figure 13 – Progress on maternal mortality: MDG 5 for developing and transition LIFDCs, 1990–2000[5]

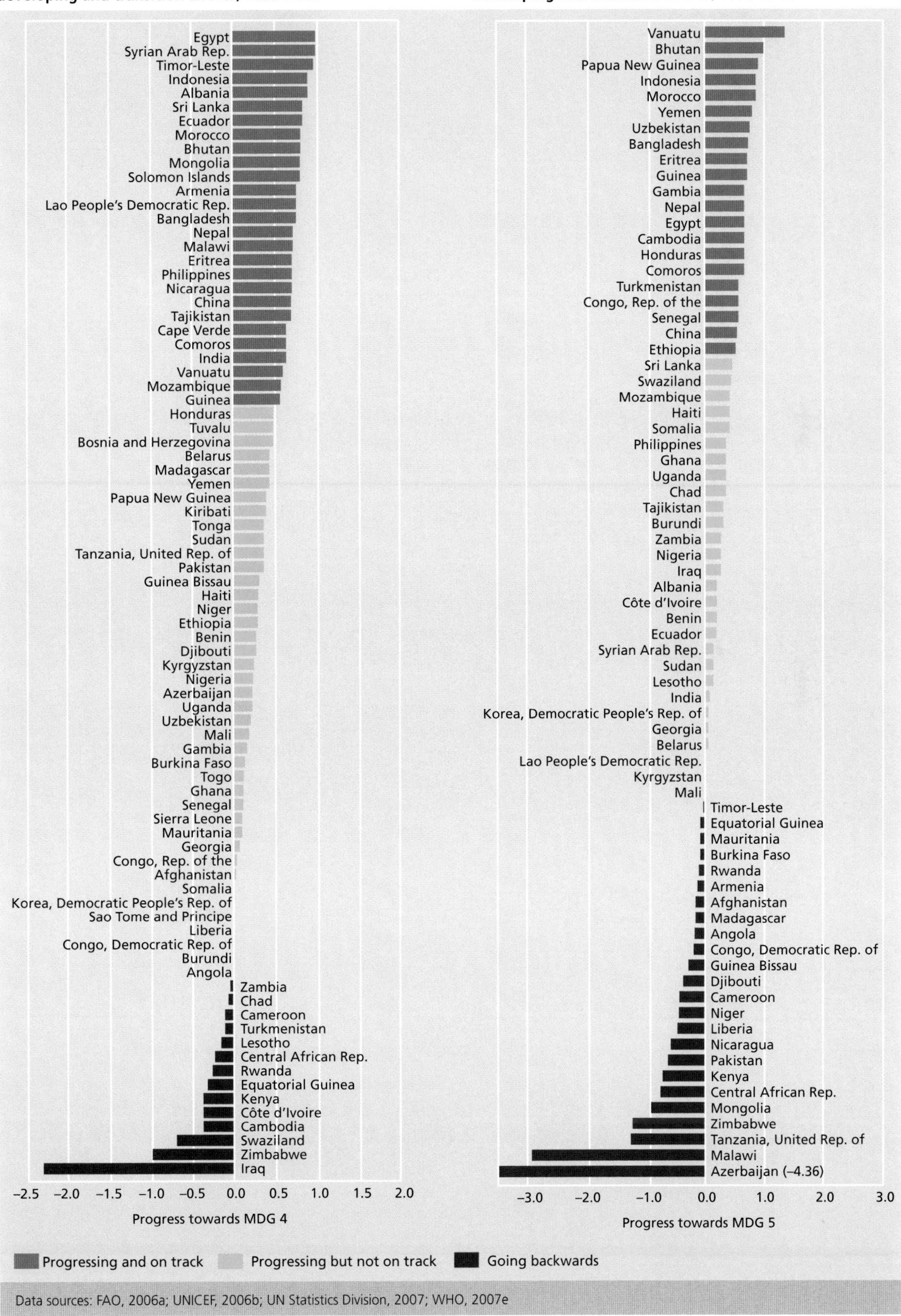

Data sources: FAO, 2006a; UNICEF, 2006b; UN Statistics Division, 2007; WHO, 2007e

1.4 Accelerating progress: making the right choices

"The poorest groups lag behind even when their countries are making progress overall. ... All interventions need to reach the poor to lessen the gap."

Disease Control Priorities Project, 2007

The MDGs commit the world to address some of the most critical issues of our time. Children depend on the decisions made today to provide them with a world in which they are given the chance to grow into healthy and productive adults, where they can contribute to economic and social development and where they can lead the next generation towards a promising future. Achieving the MDGs, or substantial progress towards meeting them, would help to lay the foundation for such a world. Unfortunately, the poorest countries are the furthest behind. Of particular concern is ensuring that overall national targets are met but not by further exclusion of the most vulnerable. Also, while it is important to increase and expand services, it is imperative that quality improves, or at a minimum does not deteriorate, in order to keep up with numeric targets.

Figure 14 – GDP, child mortality and underweight in developing and transition countries

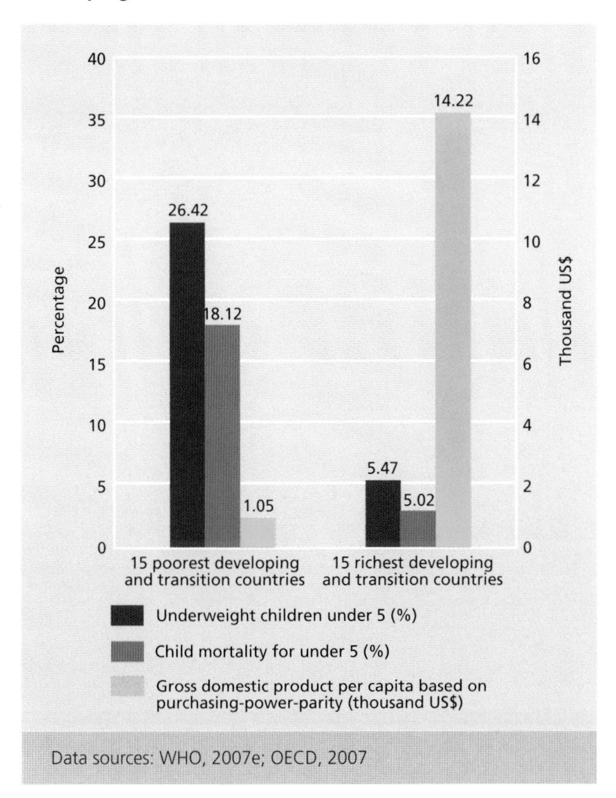

Data sources: WHO, 2007e; OECD, 2007

For a number of the MDGs, and for many countries, lack of data hinders the measurement of progress in reducing hunger and poor health. Because annual data are not available, it is difficult to say with certainty how far we need to go to meet all the MDGs. However, for the hunger goal alone, the annual rate of reduction must be accelerated to at least 26 million people per year; currently only one twelfth, or 2.1 million people escape hunger each year (FAO, 2003).

There have been some assessments of the costs of not achieving the MDGs by 2015 in terms of lost economic growth, lives not saved and insufficient reductions in child undernutrition. If the current slow rate of progress continues, the lives of 3.8 million children who could be saved will be lost every year. By 2015, 50 million children who could escape the devastating effects of undernutrition will fail to do so (UNICEF, 2006a).

Conversely, the benefits of acting are great. Reducing the prevalence of iron deficiency anaemia by means of food supplements has an exceptionally high ratio of benefits to costs – an estimated saving of US$12 billion. Similarly, the cost of averting 30 million new HIV/AIDS infections is substantial, estimated at US$27 billion. However, the potential gain from averting these infections is even greater – preventing the collapse of entire societies (Mills and Shillcutt, 2004).

Meeting the child and maternal mortality targets imply an average annual rate of reduction of 4.3 percent and 5.4 percent respectively (Disease Control Priorities Project, 2007). Progress is still slow: no substantial progress has been made in reducing child mortality in Africa, where the rate has declined by only 10 percent since 1990. Greater progress was achieved in the 1970s and 1980s.

Lofty goals to alleviate hunger and poor health, or at least meet the MDGs uniformly, can only be achieved through stronger alliances among governments, international agencies and donors, civil society, the private sector and, perhaps most important, the hungry themselves. A combination of greater budget allocations from governments and assistance from donors is required if countries are to achieve the MDGs. Solutions that could help to meet all the MDGs are known, but they are not being used as effectively as they could be, nor are governments always capable of implementing programmes on the required scale.

It is important that, within the growing alliances, donors support and work with countries to help re-allocate funds to reduce hunger, improve health, and strengthen policies and programmes to sustainably meet the MDGs. Only by prioritizing the hungry, and especially women and children, and by supporting principles of inclusion, equality, ease of access and transparency can the hungry benefit from the technological innovations that are transforming the world.

Map 3 – Inequality of hunger across the world

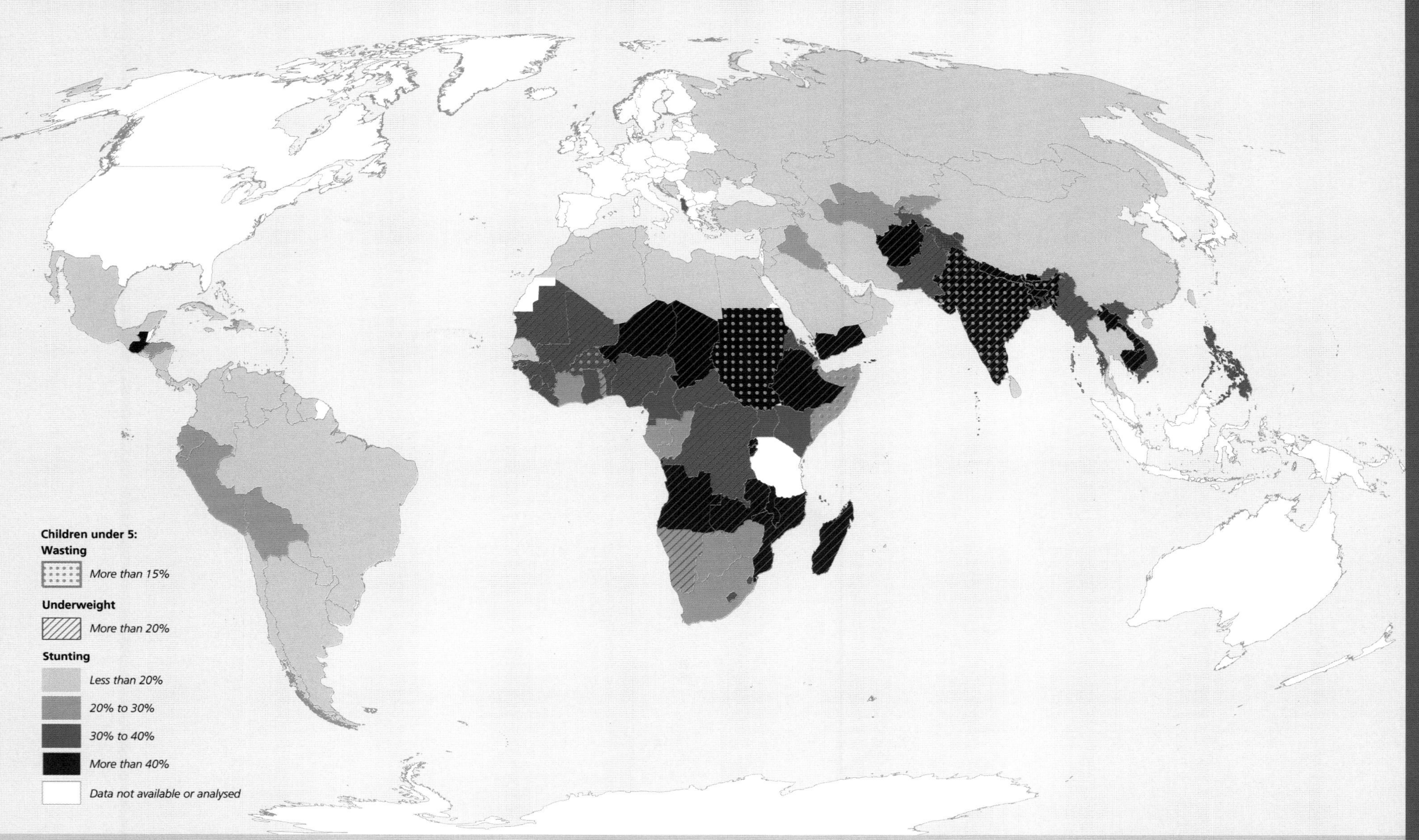

The boundaries and the designations used on this map do not imply any official endorsement or acceptance by the United Nations.
Map produced by WFP VAM.

Data sources: WHO, 2007; UNICEF, 2006

Intermezzo 3: Hunger and disease in crisis situations

Despite an increase in the number and scale of most kinds of humanitarian crises in recent years, "excess mortality" in emergencies has been falling.[6] Between 1900 and 1960, there were ten years in which the total number of disaster-related deaths exceeded 1 million each year (CRED, 2005). Armed conflicts in Bosnia and Timor-Leste produced conditions conducive to famine in the mid-1990s, but mass non-combatant mortality was averted (even if other military and moral catastrophes were not). Similarly, the serious droughts in southern Africa in 1991–1992 and 2001–2002 caused massive shortfalls in food availability, but no clearly-defined "famine" deaths ensued thanks to rapid, targeted, multi-sectoral interventions.

Underpinning such successes is an evolving awareness of the importance of micronutrients in determining how large-scale disasters unfold, and how best to respond through relief interventions. When lives are under threat, targeted actions are needed to address the most severe manifestations as rapidly as possible. There are two challenges to be met in emergency response: how to correct outbreaks of micronutrient deficiency disease at the individual level, reducing the mortality rate of those most severely affected; and how to prevent micronutrient status from deteriorating at a population-wide level, thereby preventing epidemic infection on a larger scale and linking improvement to sustainable post-crisis gains. In this sense there are two main approaches that relate to micro-nutrients: the targeted transfer of micronutrients to those in need, and empowerment with knowledge about deficiency disorders and potential solutions.

Having "something to eat" is not in itself enough to achieve a sound nutritional status, nor indeed sufficient to prevent malnutrition. For example, periods of widespread food scarcity invariably lead to scavenging and the consumption of products that are not part of a conventional diet. During the European famine of 1817, desperate Germans and Swiss ate sawdust baked into bread, carrion, their watchdogs, even grass and roots (Webb, 2002). During famine in southern Africa in 1896–1897, colonial officials reported that people were "... suffering from a disease, which I believe is caused through eating ... rotten skins, and wild roots, the effect of which causes them to purge considerably ..." (Iliffe, 1990). Such "purging" often resulted in more rapid death than if such foods had not been consumed.

Even where conventional foods are consumed, reliance on an inappropriately balanced diet can lay the foundation for morbidity and mortality. However, dietary sufficiency in energy and protein does not mean that consumption of sufficient vitamins and minerals in recommended quantities is assured. Thus, while consumption of maize and groundnuts may fill the belly and assuage hunger, such a diet is not enough to meet the full spectrum of minimum daily nutrient requirements.

Micronutrient deficiencies raise important new questions about the nature of food entitlements. "Hidden hunger", a deficiency unknown to the individual but well-known to those engaged in public action, challenges the simple concept of entitlement because "access" in this sense relates to intangibles such as the knowledge, behaviour change and microscopic inputs that are all needed to achieve sound nutrition but not demanded by those who need them the most. When they are delivered, as in emergencies, they are supply-driven, not supported by demand.

Public action to remedy and prevent outbreaks of micronutrient deficiency disorders is not only critical in emergencies: it represents the fulfilment of one of the "moral rights of the hungry" (Sen, 1997). Entitlements are secured when hungry people establish ownership over an adequate amount of food, or where their moral right to food is translated into a "practical right". Humanitarian action represents precisely that – a practical enforcement of the moral right not to die from lack of food. The humanitarian imperative demands that relief be provided unconditionally to those who are suffering, whoever and wherever they are (Webb, 2003).

That said, enforcing the right to food increasingly means delivering under extremely difficult circumstances not simply the right quantity of food, but also the right quality. Addressing vitamin and mineral deficiencies is a core aspect of humanitarian relief; it is a commitment by the international community to uphold the moral right of the hungry not just to survive, but to do so with the necessary nutrients and with dignity. But this raises questions about current limits to public action in emergencies. Delivery by humanitarian personnel of therapeutic foods in emergencies saves lives, and hence can be characterized as the "enforcement" of Sen's moral right to food, but it does so in a time-bound way. Access to micronutrient-fortified foods usually ends once an individual is released from therapeutic feeding or when an emergency relief activity winds up. Thus, the state-of-the-art use of micronutrients in saving life does little to empower the person whose life has been saved. That person typically does not gain either the knowledge about micronutrients and health needed to build effective demand for nutrients, not just food, or the ability to secure access to micronutrient-rich foods through market or other channels on a regular basis once a crisis has passed.

New approaches are needed to define ethically acceptable forms of humanitarian intervention that not only save lives but also enhance the demand for access to nutrition, not just food. Supporting the entitlements of the hungry cannot stop at ensuring access to food, because micronutrient adequacy demands access to the right foods and the right knowledge, leading to household-level behaviour change, not simply enhanced access to markets or purchasing power. Solutions to entitlement failure must therefore operate at the level of prices and markets, and equally in the domain of public health and nutrition. Solutions focused only on food quantity and not on dietary quality are likely to fall short of their intent and of our deeper responsibilities.

Excerpt contributed to the World Hunger Series *by Patrick Webb, Dean for Academic Affairs, Friedman School of Nutrition Science and Policy, Tufts University and Andrew Thorne-Lyman, Public Health Nutrition Officer, WFP.*

Milestones in the evolution of nutrition concerns in emergencies

1960s	• Responses based on available food • Foods donated determined more by availability than nutritional adequacy • Limited recognition of relevance of nutritional content of rations
1970s	• Focus on protein deficiency (in protein-energy malnutrition) • More variety in food basket, including beans, vegetable oil • Fortified blended foods (FBFs) used only in supplementary feeding
1980s	• Major agencies raise ration planning figure from 1,500 to 1,900 kcal per person per day • FBF included in most rations for completely food assistance dependent populations • Food basket increasingly based on six core items: cereals, pulses, oil, sugar, salt and FBFs
1990s	• Some agencies (including WFP) increase ration planning figure for fully food assistance dependent populations from 1,900 to 2,100 kcal • Advances in science lead to production of therapeutic foods for treating acute malnutrition (F100, F75) • Stricter limitations on the use of milk products and infant formula in crises • Requirement that internationally procured oil, salt and flour be fortified • Local production of FBFs expands in some developing countries • BP5 and high-energy protein fortified biscuits in wide use
2000s	• Greater use of local milling and fortification of cereals for relief distribution • Local (developing country) procurement of FBFs for use in third countries • Development of ready-to-use therapeutic foods for "at home" treatment of acute malnutrition • More attention to links between treatment of acute malnutrition and prevention of chronic malnutrition

Source: adapted from Toole and Waldman, 1988

Part II Undernutrition and Disease: Impacts Throughout the Life Cycle

Hunger, poor dietary intake, unsafe water, lack of hygiene and sanitation, drug-resistant disease strains and limited access to medical services continue to impede progress in controlling undernutrition and infectious diseases, despite technological advances.

Chapter 1 explores in more depth the hunger–health relationship, specifically the interaction between undernutrition and disease. **Chapter 2** traces this two-way relationship throughout the life cycle for high-burden infectious diseases and examines how undernutrition in early life has long-term effects in later life in the form of chronic disease and obesity. **Chapter 3** considers emerging threats from a hunger and health perspective, and how these threats could increase vulnerability to hunger and poor health, in particular for the most vulnerable.

"Nutrition needs to be improved to reduce the devastating effects of diseases that are keeping the hungry and malnourished from being productive members of society."

The Des Moines Declaration, 2004[7]

Throughout history the presence or absence of hunger and disease has helped to shape the course of events. For example, between 1150 and 1200, a major climatic warming throughout Europe coupled with the rise of a middle class led to improvements in diet. By the middle of the 14th century, Europe was notably overpopulated; a climate change – the so-called Little Ice Age – brought conditions that were colder and wetter than normal, resulting in decreased crop yields and drastically reduced per capita caloric intake. General health declined, and rodent and pest populations increased, a combination of circumstances that invited a major epidemic. Bubonic plague – the Black Death – swept through Europe wiping out a significant portion of the population with periodic outbreaks until the 18th century (Scott and Duncan, 2001).

Common determinants of undernutrition and disease	
Healthcare	• Lack of immunization and preventive measures such as bed nets • Limited use of chemotherapy, anti-retroviral therapy regimes • Risky behaviour
Environment	• Poor water, sanitation and personal hygiene • Crowding and dense housing conditions • Insect and pest infestations
Access to quality food	• Poor dietary diversity, energy and micronutrient intake • Food availability and production constraints • Poor nutrient absorption
Care practices	• Inadequate breastfeeding practices • Lack of adequate oral rehydration methods • Poor re-feeding diet after illness • Too few and too small meals for young children

Source: SCN, 2004a

Infectious diseases such as the plague and related mortality have been declining in developed countries over the last 200 years. But the developing world still faces daunting challenges: infectious diseases such as malaria, TB, childhood diseases including diarrhoea and pneumonia and nutritional deficiencies are all major causes of mortality and morbidity. Hunger, poor dietary intake, unsafe water, lack of hygiene and sanitation, coupled with drug-resistant disease strains and limited access to medical services continue to impede progress in controlling undernutrition and infectious diseases, despite technological advances.

The impact of undernutrition on disease

Just as hunger and health are closely related, so are undernutrition and disease. Similarly, the relationship between undernutrition and disease is bidirectional and mutually reinforcing. Undernutrition leads to a state of poor health that puts the individual at risk of infectious and chronic disease. This two-way relationship extends to the major killers today – HIV/AIDS, TB, malaria, acute respiratory infections (ARIs) and diarrhoeal disease – particularly in less developed parts of the world.

Undernutrition is a physical manifestation of hunger, as disease is of poor health.

The hungry are less able to fight disease than well-fed people. Undernutrition contributes directly to disease by depressing the immune system and allowing pathogens to colonize, further depleting the body of essential nutrients. Lack of energy and/or micronutrients compromises the immune system, which in turn increases susceptibility to infectious and chronic diseases.

Undernutrition causes a general weakening of the body that can lead to more frequent incidence of infectious disease and/or increased severity and duration of the disease. Evidence suggests that even mild forms of undernutrition produce adverse effects on the immune system, particularly among children and pregnant women. As a consequence, poor nutritional status exacerbates the progress of infection

Figure 15 – Undernutrition and disease determinants

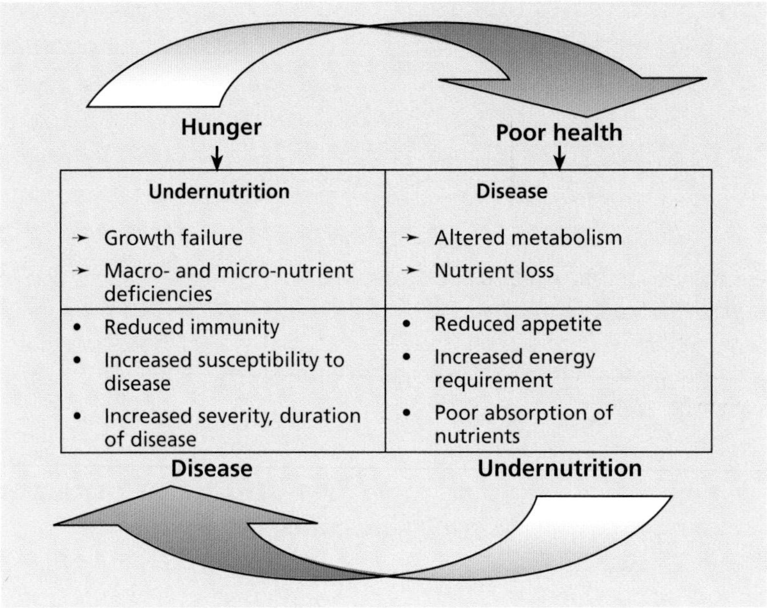

Hunger → **Undernutrition**	Poor health → **Disease**
→ Growth failure → Macro- and micro-nutrient deficiencies	→ Altered metabolism → Nutrient loss
• Reduced immunity • Increased susceptibility to disease • Increased severity, duration of disease	• Reduced appetite • Increased energy requirement • Poor absorption of nutrients
Disease	**Undernutrition**

and increases the effects and duration of disease (Tomkins and Watson, 1989).

Simply put, an undernourished child tends to suffer more days of sickness than a well-nourished child. A number of studies have confirmed the strong relationship between initial nutritional status and the duration and severity of infectious disease (Pelletier et al., 1995; Schorling et al., 1990).

Moderate and severe forms of undernutrition also undermine the ability to both resist and recover from infectious disease. A prolonged disease in an undernourished young child can lead to growth failure and even death. Until recently, child mortality was mainly attributed to disease, with little attention given to the relationship with undernutrition. Concurrent undernutrition and infection produce an interaction that is more serious than if one of the two were working independently. More children in the developing world die from the simultaneous presence of undernutrition and disease than from hunger or illness alone (Tomkins and Watson, 1989).

Nutritional diseases

Physical growth failure resulting from undernutrition is captured through various anthropometric measures, for example stunting and underweight (SCN, 1990). Undernutrition can become visible in nutritional diseases such as *kwashiorkor* and *marasmus* (Williams et al., 2003).

Unfortunately, despite the broad acceptance of the causal relationship between undernutrition, disease and child mortality, programmes and resources have disproportionately been directed toward managing infectious diseases rather than preventing hunger and undernutrition (Schroeder, 2001).

Disease's impact on undernutrition

"Infections, no matter how mild, have adverse effects on nutritional status. The significance of these effects depends on the previous nutritional status of the individual, the nature and duration of the infection, and the diet during the recovery period. Conversely, almost any nutrient deficiency, if sufficiently severe, will impair resistance to infection" (Scrimshaw and San Giovanni, 1997).

Acute and chronic infections can have serious impacts on nutritional status, triggering different reactions, including reduced appetite and impaired nutrient

Disease and undernutrition: findings from the Liberia survey

- An alarming proportion of Liberian children cannot reach their full potential as a result of undernutrition. Malaria and diarrhoea are the most common causes of child morbidity and mortality. Overall, 36 percent of deaths among children under 5 were attributed to malaria; 19 percent were attributed to diarrhoea.
- Underweight is primarily associated with low birth size and episodes of illness, in particular fever and diarrhoea. Children who are smaller than normal at birth and who have recently experienced fever or diarrhoea are significantly more likely to be underweight.
- Wasting levels are associated with healthcare and health indicators; children who had not received a measles immunization or deworming treatment and children who had recently experienced fevers, coughing with shallow breaths or diarrhoea were significantly more wasted.
- Low birthweight is associated with the nutritional status of women. Undernourished mothers were 31 percent more likely to have low birthweight babies compared with nourished women. Undernourished women were less likely than well-nourished women to have babies of above normal birthweight.

Note: see the box on page 38 for background to this survey.
Source: WFP, 2006

absorption. Even when nutrients are absorbed, they may still be lost as a result of the infection. The magnitude of the loss is generally related to the severity of the infection, and associated with:

- poor absorption of nutrients;
- increased energy requirements; and
- decreased appetite resulting in weight loss or failure to gain weight.

These conditions are particularly common and problematic in sick children and can result in growth failure. Weight loss and underweight status are considered to be a precursor to wasting and acute undernutrition.

Disease becomes more severe as the effects of undernutrition and disease repeat in a vicious cycle. Findings from community studies in Guatemala, for example, indicate that children with ARIs or diarrhoea consume fewer calories per day than children without such infections. Also, more severe infections can lead to much larger deficits in food and caloric intake. African children examined during the acute phase of measles consumed 75 percent fewer total calories than they did after recovery (Stephensen, 1999). Poor families have less opportunity to obtain access to quality food, health services and medicines – hence illnesses may last longer. It is important to highlight that when sufficient quality and quantity of food is available, illnesses do not cause high prevalence of undernutrition in children (Lutter et al., 1992). However, illness exacerbates undernutrition in a poorly nourished child, and the worse the diet the more undernourished a child becomes.

In many poor food-insecure households, where the undernutrition disease cycle can often be repeated for a prolonged period of time, the vicious effects seem insurmountable and inter-generational.

"Momentum is building, but disease is still way out in front. The numbers are so big that they can numb us into indifference: 5,000 people dying every day from tuberculosis, 1 million dying every year from malaria. Behind each of these statistics is someone's daughter, someone's son, a mother, a father, a sister, a brother."

Paul David Hewson (Bono), 2005

Infectious diseases in childhood

Each year, nearly 10 million children die before reaching their fifth birthday. Most of this mortality is attributable to three diseases: diarrhoeal disease, ARIs (including pneumonia) and malaria. The duration and severity of these infectious diseases are compounded by undernutrition.

Approximately half of all deaths in children under 5 are a direct result of hunger. These 5.6 million annual deaths occur in children with common infectious diseases, particularly diarrhoea, acute respiratory infections and measles, which generally would not be fatal in well-nourished children (Bryce et al., 2005). Nearly 800,000 of these deaths are due to concurrent vitamin A deficiency in areas where this deficiency is prevalent (Rice et al., 2004).

The estimated proportions of deaths in which undernutrition is a synergistic cause are roughly similar for diarrhoea (61 percent), malaria (57 percent), pneumonia (52 percent), and measles (45 percent). Synergistic deaths do not occur unless both conditions, undernutrition and infectious disease, are present (Bryce et al., 2005). Both are primary causes of childhood deaths.

Figure 16 – Major causes of death among children under 5 worldwide

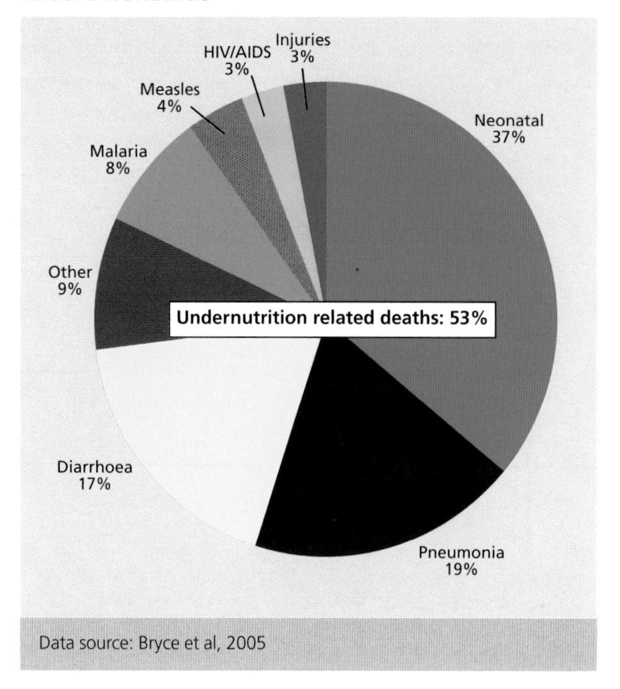

Data source: Bryce et al, 2005

Children adapt to low macronutrient and micronutrient intake by reducing their physical activity; at the same time, their growth slows. If inadequate intake of food persists, their immune systems are compromised and they become more vulnerable to disease and infection. Numerous studies have established a strong association between undernutrition and increased morbidity and mortality for a range of common childhood diseases, including ARIs and diarrhoeal disease.

Diarrhoeal disease. Diarrhoeal disease is a leading cause of childhood morbidity and mortality in many developing countries. Repeated episodes of diarrhoea in children can lead to undernutrition, poor growth, decreased immunity, and in many cases death. Episodes of diarrhoea in undernourished children tend

BURDEN OF CHILDHOOD DISEASES

Diarrhoeal disease	ARIs
6,000 children die of water-related diseases every day. Over **4 billion** cases of diarrhoeal disease cause **2.2 million** deaths every year, mostly among children under 5.	Pneumonia – a severe form of ARI – contributes to **19 percent** of all deaths among children under 5 and **26 percent** of neonatal deaths.
Source: UNICEF, 2007a	Source: UNICEF and WHO, 2006

Childhood disease	Risks and effects	Contributing factors
Diarrhoeal disease	• Reduced intestinal absorption • Deficiencies of vitamin A and zinc	• Undernutrition including micronutrient deficiencies • Limited access to clean water • Poor sanitation • Contaminated baby bottles • Early introduction to solid foods • Improper food storage
ARIs	• Low birthweight • Deficiencies of vitamin A, zinc, vitamin D and calcium • Increased susceptibility to other diseases	• Undernutrition including micronutrient deficiencies • Prior respiratory infections • Lack of immunization • Presence of HIV/AIDS • Tobacco smoke; air pollution

Source: adapted from UNICEF, 2002

to be more severe and of longer duration, suggesting a vicious cycle of diarrhoea and undernutrition. Diarrhoeal disease can reduce energy consumption by as much as 15–20 percent in children.

Poor nutritional status including deficiencies of vitamin A and zinc can increase the severity of diarrhoeal disease. A number of studies have shown that continuing diarrhoea in very young children negatively affects their growth (Checkley et al., 2003; Assis et al., 2005).

Children are undernourished in poor communities because they do not get enough food and not because they suffer from diarrhoea. Ensuring that deprived children have enough to eat still seems the best approach to alleviate the problem of undernourishment (Briend et al., 1989).

Acute respiratory infections. ARIs are among the most common diseases in infants and young children in developing countries. An important contributing factor linked with ARI deaths is undernutrition. In developing countries, the percentage of children under 5 affected by a form of ARI increased from 16 percent in 1990 to 54 percent in 2005 (UNICEF, 2006b). The increase is related to a number of factors: severe undernutrition, opportunistic diseases such as HIV/AIDS, and environmental factors including tobacco smoke and indoor air pollution. Children under 2 have the highest incidence of ARIs.

Evidence from Brazil

A nine-year study carried out in north-eastern Brazil examined the magnitude and duration of the association of early childhood diarrhoeal infections with growth faltering during childhood. The evidence showed that early childhood diarrhoea and subsequent intestinal infections affected the cognitive and physical growth of children (Moore et al., 2001).

Micronutrients status has been implicated as a risk factor in ARIs. For example, respiratory diseases have been associated with an increased risk of developing vitamin A deficiency; in turn, vitamin A deficiency has been shown to increase the severity of respiratory disease, especially among school-age children. Nutrient supplementation and sustained breastfeeding appear to reduce ARIs. Zinc supplementation also appears to have some potential in reducing morbidity and mortality related to ARIs.

Knowledge gaps: nutrition and childhood diseases

Overall, there is a striking gap in information regarding the burden of childhood diseases, especially in the poorest countries with the highest rates of child mortality. The duration of a diarrhoeal episode and its nutritional effect have not been captured well in children's growth profiles.

Micronutrient deficiencies are strongly associated with ARIs and diarrhoeal disease. Additional research is needed to delineate potential beneficial and possible

Map 4 – Mortality and childhood diseases

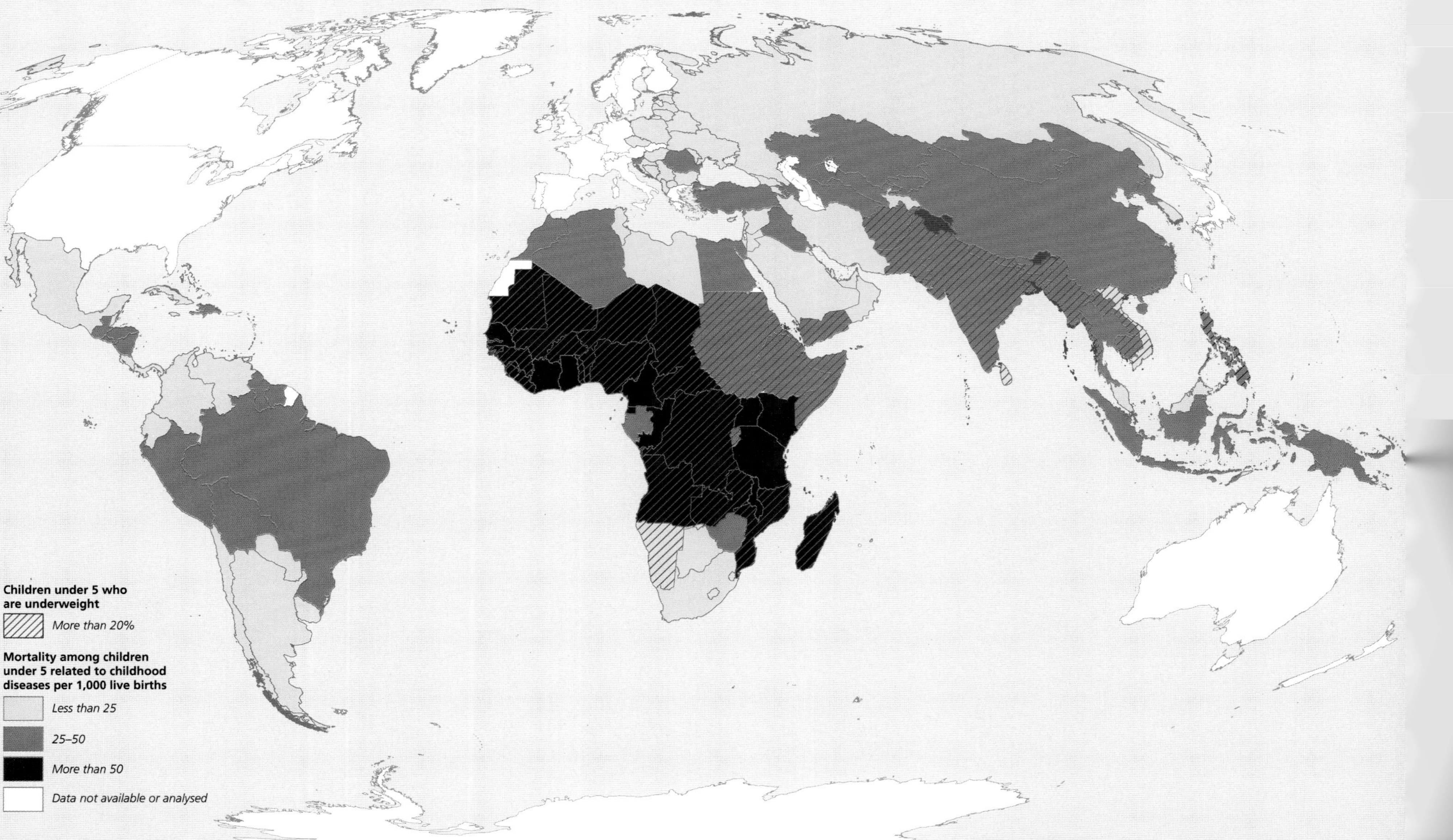

The boundaries and the designations used on this map do not imply any official endorsement or acceptance by the United Nations.
Map produced by WFP VAM.

Data source: WHO, 2007

wfp vam
vulnerability analysis and mapping

ARIs and micronutrient deficiency

Undernourished children are twice as likely to develop ARIs, particularly when they are not breastfed and suffer from a low socio-economic status (Deb, 1998). Evidence shows that ARIs interfere with the absorption process of essential micronutrients, including vitamin A. Several studies report that a large proportion of infants remain vitamin A deficient even after a substantial dosage of vitamin A supplementation, due to frequent and recurring respiratory infections, particularly those accompanied by fever (Rahman et al., 1996).

detrimental interactions between nutritional status and micronutrient supplementation in children with these diseases.

Infectious diseases that cut across the life cycle

Malaria and undernutrition strongly interact; nearly 57 percent of malaria deaths are attributable to undernutrition (SCN, 2004b).

Burden of malaria

- **Over 500 million** people are infected every year.
- **1 million** children die of malaria every year, mostly in Africa.
- **90 percent** of mortality occurs in Africa.

Source: UNICEF, 2007b

Malaria remains a pervasive public health problem throughout the tropics, affecting children, adolescents, adults and the elderly in over 100 countries and territories. Over 40 percent of the world's population is at risk of contracting the disease. The disease has its greatest impact among children under 5 and pregnant women. It has been found that when someone, usually a child, is already severely undernourished and contracts malaria they have a much higher risk of dying.

Malaria is a significant contributor to the vicious cycle of food insecurity, chronic hunger and poor health. Malaria consistently undermines gains in family food security, as it reduces the productivity of income earners, draws on income and savings to treat the disease and, ultimately, causes loss of life – especially in young children, but also in productive family members.

Undernutrition and malaria: a close relationship

There is ample, though complex, evidence to show an important relation between undernutrition and malaria. Several studies have examined the association – whether antagonistic, synergistic or neutral – between different types of undernutrition, including protein-energy deficiencies, stunting and underweight. In addition, there are a number of studies with conclusive evidence showing the interaction between undernutrition and *Plasmodium falciparum* (Shankar, 2000).

Malaria

Life cycle stage	Hunger/health risk
Pregnancy	• Increased risk of morbidity, anaemia, stillbirths, and congenital infections. • Increased risk of pregnancy complications.
Infants and young children	• Low birthweight babies. • Compromised iron intake further increases risk of anaemia.
School-age children, adolescents, adults and elderly people	• Malaria-infected people have increased risk of severe anaemia.

These studies generally provide evidence that malaria appears to exacerbate undernutrition. Once a person is infected, malaria can result in weight loss, undernutrition and growth failure.

A number of studies in Africa have reported weight loss in young children following a malaria attack (Shankar, 2000):

- The Gambia: a study noted that malaria was significantly related to lower weight gain and growth faltering in children under 3.
- Chad: malaria patients were 0.57 times as likely to be undernourished, and undernourished malaria patients were 1.5 times more likely to die.
- Burkina Faso: underweight children were 0.9 times as likely to be infected.

Malaria and undernutrition: a complex relationship

- Two studies carried out in Brazil and the former Union of Soviet Socialist Republics reported greater frequency of more severe malaria in those who were undernourished (Shankar, 2000).
- A small-scale study in India reported an increase in parasite density with improved nutritional status, indicating that undernutrition was protective.

- Zambia: underweight children were 1.27 times more likely to be infected.
- The Sudan: stunted children were 1.4 times more likely to have had a recent malaria illness.

The synthesized data may not give the complete picture, but they indicate a complex two-way relationship between malaria and undernutrition: undernutrition and malaria aggravate each other and increase the likelihood of death.

General improvements in dietary intake, through improved access to quality food, in particular for young children, are likely to have a large impact on reducing the burden of this disease.

Malaria and micronutrient deficiencies

Individuals with malaria have lower concentrations of several micronutrients than people who are not infected. Consequently, the prevention and treatment of malaria is closely linked to the interaction between specific micronutrients and disease attacks.

Micronutrient deficiencies also play a role in controlling malaria morbidity and mortality. Studies show promising results regarding the role of micronutrient supplementation – particularly vitamin A and zinc – in reducing morbidity in children living in malaria endemic areas.

Knowledge gaps: nutrition and malaria

More detailed research is needed to clarify the links between specific nutrients and malaria morbidity, particularly the effect of vitamins and minerals – vitamin A, zinc and iron. Studies reported reduced malaria morbidity associated with vitamin A and zinc supplementation in pre-school children, but such effects were not seen in other groups living in areas where malaria is endemic.

More research is still needed on the positive and negative effects of iron supplementation in malaria endemic areas.

HIV/AIDS – a tremendous burden. The dual burden of HIV/AIDS and hunger threatens agriculture and rural development, reducing productivity and income, and increasing expenditures on medical treatment. The dependency ratio in families as a result of rising numbers of dependents relying on a smaller number of income earners also grows. The disease precipitates the loss of traditional farming methods, inter-generational knowledge and specialized skills, practices and customs (FAO, 2003). Thus hunger closely accompanies HIV/AIDS.

Figure 17 – The two-way relationship: malaria and micronutrient deficiencies

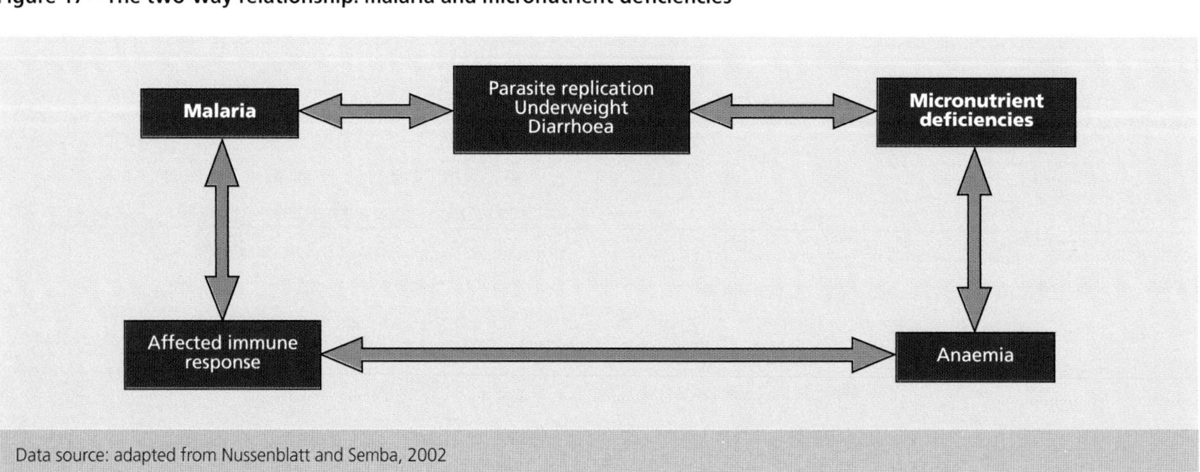

Data source: adapted from Nussenblatt and Semba, 2002

Burden of HIV/AIDS

Number of people living with HIV in 2006:
Adults: **37.2 million**
Women: **17.7 million**
Children: **2.3 million**

Number of AIDS deaths in 2006:
Adults: **2.6 million**
Children under 15: **380,000**

Source: UNAIDS, 2006

Hunger contributes to HIV in that it:

- increases migration in search of food and income;
- increases risk taking for food and income; and
- exacerbates gender inequality – when there is limited food in the household, women often suffer first and have to seek food elsewhere.

Increased energy needs and HIV

People living with HIV (PLHIV) have special nutritional needs, even though the clinical course of HIV infection may be highly variable. The relation between HIV and undernutrition becomes evident with the early clinical signs of the disease. The intestinal track is affected, resulting in poor absorption and loss of appetite, reducing the quantity of food consumed.

Wasting in adults and growth failure in children, the hallmarks of AIDS, are strong predictors of morbidity and mortality in infected individuals. Studies have shown that energy demands are increased in PLHIV: an increase in resting energy expenditure has been observed in HIV patients. Dietary intake is compromised as a result of poor appetite, and lack of access to sufficient quality food can cause further nutritional decline.

The target is to maintain growth in children and weight in adults. WHO recommends that asymptomatic HIV-positive adults and children increase their energy intake by 10 percent to maintain body weight and ensure growth. Among HIV symptomatic adults, energy needs increase by 20–30 percent; children living with HIV experiencing weight loss require an energy increase of 50–100 percent (WHO, 2003).

Micronutrient deficiencies and HIV

Micronutrient deficiencies are known to impair immune function, which is likely to increase the risk of disease transmission and progression (Villamor et al., 2004; Fawzi et al., 2005; Kupka et al., 2005). As with macronutrients, HIV infection by itself is likely to affect nutrient absorption and contribute to micronutrient deficiencies.

Studies have consistently shown that multivitamin supplementation – vitamins B, C and E – in HIV-infected pregnant women leads to a reduction in the incidence of adverse pregnancy outcomes such as low birthweight and premature birth. These vitamins are also beneficial in HIV-infected men and women before

HIV/AIDS

Life cycle stage	Hunger/health risk
Pregnancy	• Increased risk of undernutrition because of recurring infections. • Decreased nutrient stores and poor nutrient absorption.
Infants and young children	• Low birthweight; significant growth failure. • Increased risk of mother-to-child transmission through breastfeeding.
School-age children and adolescents	• Increased risk of stunting and wasting. • Increased energy needs resulting from poor absorption. • Increased risk of transmission and infection through sexual activity.
Adults and elderly people	• Increased risk of wasting. • Increased energy requirements. Poor nutrient absorption.

Source: Fawzi and Mehta, 2007

Map 5 – The burden of malaria across the world

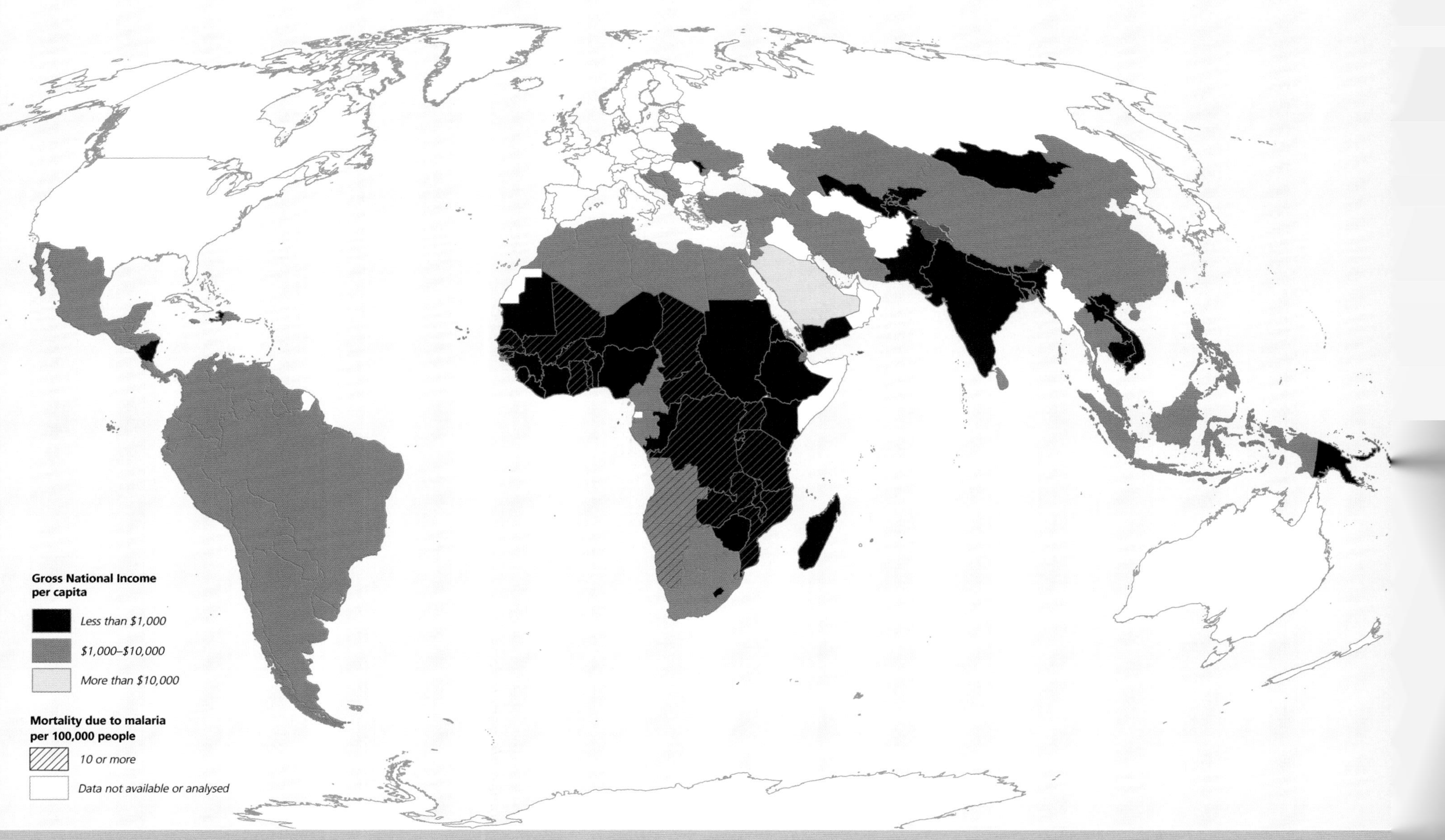

The boundaries and the designations used on this map do not imply any official endorsement or acceptance by the United Nations.
Map produced by WFP VAM.

Data sources: OECD, 2006; WHO, 2007

starting ART as they may slow disease progression and reduce mortality (Fawzi and Mehta, 2007).

One treatment strategy is to focus on improving the nutritional status of individuals already infected with the disease. Another strategy is to ensure food security in populations at risk of infection. Understanding these linkages facilitates care and treatment once the disease takes hold; it is also critically important in prevention and mitigation.

It is generally accepted that ART is most appropriately delivered as part of an integrated and coordinated strategy. Adequate nutrition helps reduce side-effects and improves tolerance to drugs, especially at the initial stages of treatment. Practitioners are beginning to recognize the important role of a short-term nutrition/food package to accompany the first 6–12 months of treatment: the package helps patients to adjust to the medication and to counter wasting (Greenblott, 2007). Food support that improves an individual's dietary intake helps to minimize the side-effects of ART and enhances adherence to and the effectiveness of treatment. Vitally, well-nourished individuals are less likely to die in a given period.

Food-based approaches to increase energy intake and to diversify diets should be encouraged in order to meet increased energy requirements, increase appetite and to reduce the risk of micronutrient deficiencies in particularly vulnerable populations.

Evidence from Zambia

A recent study in Zambia analysed the effects of giving micronutrient-fortified food to HIV-positive people who received monthly household food rations provided by WFP. Results indicated that a monthly household food ration for food-insecure patients starting ART would improve adherence (Megazzini et al., 2006).

Breastfeeding and HIV

Breastfeeding accounts for more than a third of mother-to-child transmission (MTCT) of HIV in sub-Saharan Africa. The transmission rate varies between 15 percent and 40 percent in untreated cases, but it can be contained at a level as low as 2 percent when preventive measures are taken. For example, nutrition-related practices, including exclusive breastfeeding, become critical factors in child survival (Coutsoudis et al., 2001).

The Joint United Nations Programme on HIV/AIDS (UNAIDS) recommends that breastfeeding in HIV-infected women is to be avoided when replacement feeding is acceptable, feasible, affordable, sustainable and safe. However, these conditions are rarely met in less developed regions of the world. Total avoidance of breastfeeding is usually not an option because of the high cost of formula, inadequate hygiene conditions, the greater risk of morbidity among non-breastfed infants, and the social unacceptability of such a practice.

If replacement feeding as outlined by UNAIDS is not available, it is recommended that HIV-infected women exclusively breastfeed their infants for the first few months, followed by a complete discontinuation of breastfeeding as soon as possible: the risk of HIV transmission is much greater for infants that are partially breastfed than for those that are exclusively breastfed.

Knowledge gaps: nutrition and HIV

The link between HIV infection and macronutrient requirements is a pressing area for further research. The current evidence is insufficient to recommend a change in the composition of a typically balanced diet, for example increasing the protein or the fat fraction of the diet for PLHIV.

There is mounting evidence that supplementation with some micronutrients, particularly vitamin A/beta carotin, may have adverse outcomes in PLHIV, especially pregnant women. A baseline assessment of nutritional status with reference to such nutrients is recommended before prescribing supplementation beyond the recommended dietary allowance.

Map 6 – HIV/AIDS mortality in children under 5

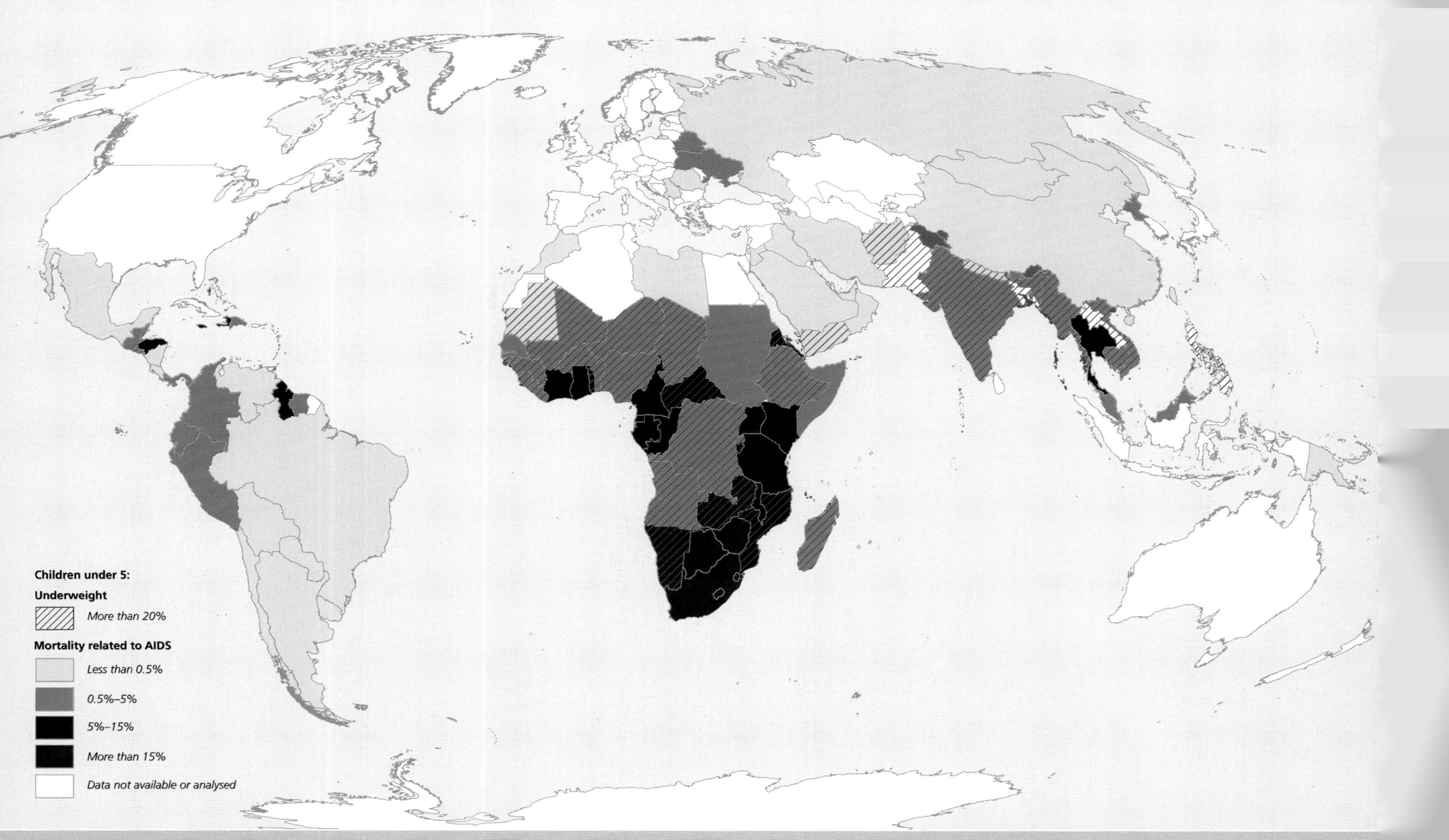

The boundaries and the designations used on this map do not imply any official endorsement or acceptance by the United Nations.
Map produced by WFP VAM.

Data source: WHO, 2007

Periodic vitamin A supplementation in children with HIV, however, appears to have similar benefits as in HIV-negative populations – for example the reduction in mortality and morbidity from illnesses such as diarrhoea – and is recommended for children older than 6 months.

There are concerns about the safety of routine iron supplementation. However, data are limited and until further evidence is available no change is recommended in routine iron and folate supplementation as suggested by WHO for pregnant women in HIV-infected populations.

Tuberculosis (TB) – still a heavy burden. Like HIV/AIDS, TB has a tremendous impact on livelihood and food security, partly because a large number of people who suffer from TB are also infected with HIV.

TB is a common infectious disease; almost 9 million new cases occur each year worldwide. One third of the world's population is infected with *Mycobacterium tuberculosis* – a large portion of whom are at risk of developing the clinical disease. Once a person has been infected the disease usually remains latent, with a 10 percent lifetime risk of the clinical disease developing. This risk is higher in children, PLHIV, individuals with cancer, undernourished people and the elderly (Shah et al., 2001).

Burden of TB (in 2005)

- **8.8 million** new cases
- **1.6 million** people died
- **195,000** TB cases were associated with HIV infection

Sources: WHO, 2007a, b

The association between hunger and TB is well recognized, with poor nutrition increasing the risk of TB and the development of TB often resulting in poor nutritional status (Cegielski and McMurray, 2004).

A number of studies have found a correlation between undernutrition and TB (Chatterjee et al., 1968; Harries et al., 1985, 1988; Onwubalili, 1988; Scalcini et al., 1991; Cegielski and McMurray, 2004). However, it is difficult to establish cause and effect – the role of nutrition in TB cannot be isolated from possible confounding factors, e.g. poor hygiene, lack of adequate medical care, overcrowding and poor housing. It is equally plausible that TB causes undernutrition rather than that undernutrition leads to increased risk or severity of TB. More clinical studies are needed to better understand the potential impact of food and nutrition interventions on risk of TB and on the outcomes of TB treatment.

Figure 18 – Estimated burden of TB, 1990 and 2005

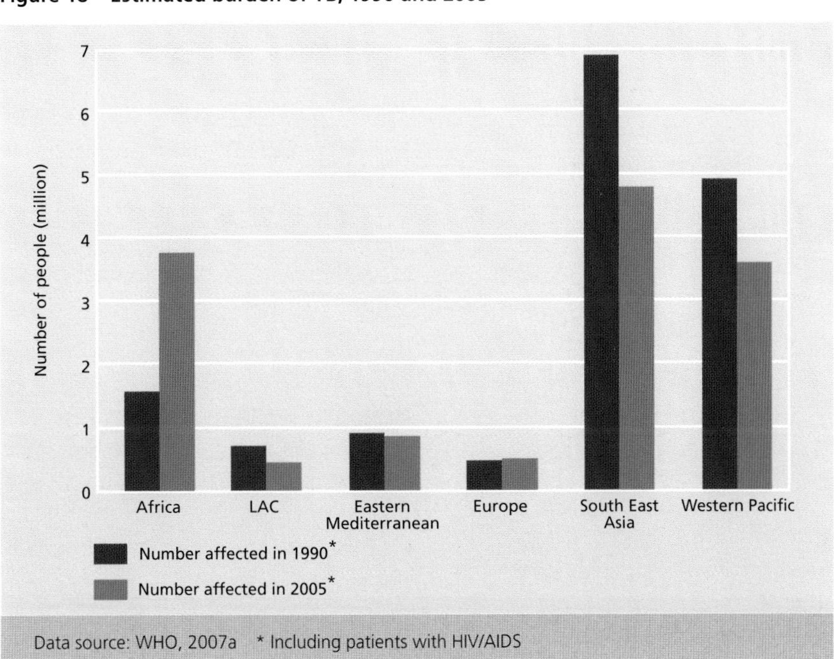

Intermezzo 4: AIDS and hunger – challenges and responses

As the multi-faceted impact of the AIDS hyper-epidemics continues to intensify across southern Africa, the world is witnessing the development of dangerous new interactions that threaten national social and economic development. Recognition of the complex, long-wave nature of the AIDS epidemic – and how it is fundamentally driven by people's livelihood strategies, constraints and opportunities – has been slow coming.

Examining the interaction between AIDS and hunger, we can see several vicious cycles. One of them, undernutrition and disease, gets to work in the human body immediately after an individual has become infected with HIV. Another vicious cycle revolves around a household's degree of access to the food it needs. On the upstream side of viral transmission, food insecurity may put poor people at greater risk of being exposed to HIV, for example through forced migration to find work or through poverty-driven adoption of transactional sex as a survival strategy. On the downstream side, the various impacts of chronic illness and premature mortality on household assets and resources are well documented. Throughout, it is the poor, and especially poor women, who are least able to cope.

The primary livelihood base of most people affected by HIV/AIDS is agriculture. In eastern and southern Africa, the AIDS epidemic is already having serious consequences for agriculture by affecting adults at the height of their productive years, making it difficult for the poor – and again especially poor women – to provide food for their families. The critical constraint, however, is often not labour in itself: it may be lack of cash resulting from the new financial demands brought by the illness. Some studies show a relatively modest impact on agriculture, but it is important to realize that this may change when young adults, who are disproportionately at risk from HIV, become household heads. We know very little of the long-term impacts of AIDS in terms of fracturing the transfer of knowledge and skills from farmers to their children.

The AIDS crisis is essentially multi-sectoral in cause and consequence. Responses that match the size of the problem are overwhelmingly sectoral and vertical, with a few, mostly small-scale innovations. Rhetoric abounds, but most development organizations remain locked in comfortable systems and timelines that simply do not align with the dynamics of AIDS. Against the backdrop of the epidemic, it is easy to succumb to a convenient state of denial or a creeping sense of professional paralysis.

To an agriculture professional who asks "Why should I bother about AIDS?" the answer is simple: if the agricultural sector in Africa fails to take HIV/AIDS into account proactively, it will not achieve its primary objective of improving food production and access. Similarly, international agricultural organizations supporting countries in Africa need to mainstream AIDS to remain relevant; otherwise the first MDG – eradicating extreme poverty and hunger – will be just a dream.

However, there is a sign of hope: communities are responding, community-based, non-governmental organizations are actively innovating, and governments are beginning to go beyond declarations on paper to put in place AIDS-responsive programmes. The recent conferences organized by the International Food Policy Research Institute in Durban in 2005 and the Africa Forum organized by Project Concern International in Lusaka in 2006 not only demonstrated the increasing demand for knowledge, but also the plethora of grassroots innovations aimed at stemming the "dual epidemics" of AIDS and hunger.

Applying a context-specific "HIV lens" to food and nutrition programming often reveals simple modifications to programmes that can improve their impacts on both hunger and AIDS. Vicious cycles can be broken and become virtuous. Improving rural livelihoods and agricultural production can help to reduce the spread of HIV and the impacts of AIDS. Programmes that reduce the need for poor people to migrate to look for work, for example by restoring degraded land, can reduce their risk of being exposed to the virus.

Applying a "food and nutrition lens" to AIDS programme strategies of prevention, care and treatment can likewise reveal potential synergies. For example, individuals who are undernourished when they start ART are six times more likely to die in a given period than well-nourished people living with HIV. We know that good nutrition improves the efficacy of the drugs, minimizes the side-effects and enhances adherence to treatment regimens. Such effects will generate short-term benefits for the patient and major long-term benefits, because better adherence slows the development of drug-resistant strains. But responses have to go well beyond short-term food assistance: they must be founded on the link between agriculture and health programming, with livelihood security as the pivotal interface.

Overriding all this new activity, however, is the clear and present need to increase the scale of effective responses to match the scale of the epidemic in order to generate a real, sustained impact. At the same time, it is important to identify which community-driven responses are working before looking at ways to enhance them and to provide additional nutritional support where local response capacity is exceeded. Governments and international organizations need to work together to develop strategies for simultaneously strengthening community resilience and creating synergistic forms of state-led social protection.

"Learning by doing" is a well-worn adage; it is a useful response when time is constrained. But for any "doing" to be accompanied by real learning, simple actionable systems of monitoring and evaluation – woefully neglected in the past – need to be put in place. For the knowledge from learning to be effectively disseminated, stakeholders must become better connected. Networks such as RENEWAL have sprung up in recent years to catalyse and scale up such learning in real time.

In 2003, the United Nations recognized the "triple threat" of AIDS, food insecurity and diminishing capacity in Africa; in June 2006, Article 28 of the United Nations General Assembly's Political Declaration on AIDS explicitly called for "all people at all times to have access to sufficient, safe and nutritious food ... as part of a comprehensive response to HIV/AIDS". We have the evidence and a mandate – what we need now is more effective action.

Contributed to the World Hunger Series *by Stuart Gillespie, Senior Research Fellow, International Food Policy Research Institute and Director, Regional Network on AIDS, Livelihoods and Food Security (RENEWAL).*

Increased nutrition needs with TB

Irrespective of the causes, undernutrition and TB are closely related. As with HIV/AIDS, TB is associated with severe wasting, which is probably a result of factors such as decreased appetite and increased nutrient losses, altered metabolism and reduced absorption of nutrients (Macallan et al., 1998; Paton et al., 1999, 2003). Loss of weight in TB patients is a strong predictor of mortality (Mitnick et al., 2003; Zachariah et al., 2002).

Early evidence

A notable study was conducted among British and Russian prisoners of war held in German camps during the Second World War. The prisoners shared the same conditions, but the British received Red Cross food supplements amounting to 30 g of protein and 1,000 kcal per day. The British prisoners were observed to have a markedly lower TB rate, 1.2 percent compared with 15–19 percent for the Russian prisoners (Leyton, 1946).

In one trial, patients who had started an anti-TB treatment in the previous two weeks were randomized to receive either standard nutritional counselling or counselling with high-energy supplements for six weeks. Patients in the supplemented group had a significantly greater increase in body weight, total lean mass and grip strength at the end of the study (Paton et al., 2004).

Knowledge gaps: nutrition and TB

Although nutritional status seems to be related to the risk of developing active TB in individuals who have latent infection, it is still unclear whether any type of nutritional intervention will help to prevent infected individuals from developing the active disease. More studies are needed to improve understanding of the relationship between macronutrients and TB. It is also important to know the impact of nutritional interventions (macronutrients and micronutrients) on the outcomes of TB treatment, and to understand more clearly how the health and nutrition sectors can collaborate in filling the knowledge gaps.

The benefits of supplementation with vitamin A, zinc or multivitamins in TB patients have been ambiguous to date. Supplementation with vitamin D may improve TB cure rates as well as decrease susceptibility to infection, but further research is needed. For this reason, the supplementation of any micronutrient beyond the recommended dietary allowance is not recommended in patients with TB. More randomized trials with large sample sizes, standardized doses of micronutrients and longer duration of follow-up are needed to determine more precisely the role of micronutrients in TB treatment.

Although the association between undernutrition and tuberculosis is not contested, filling these important knowledge gaps should be a public health priority. Also, better evidence is needed to establish how nutritional interventions can improve health outcomes for TB patients.

Intermezzo 5: Food support and the treatment of tuberculosis

TB is often exacerbated by poverty and hunger, particularly in countries with complex emergencies. For example, Afghanistan and Somalia have long suffered from civil strife and a very high TB burden. In Afghanistan, which has a population of 30 million, 50,000 people develop TB and 10,000 die of TB every year. In Somalia, with a population of 9 million, 20,000 people develop TB and 4,000 people die of TB every year. These are among the highest incidence rates of TB in countries that have yet to be seriously affected by the HIV/AIDS pandemic.

The good news is that TB care is technically quite well established. The World Health Organization developed a TB control strategy called DOTS – directly observed treatment, short-course – in the 1990s (WHO, 2007). The DOTS strategy is composed of effective diagnostic and treatment procedures: diagnosis and treatment with six to eight months of chemotherapy with multiple anti-TB drugs. DOTS has recently been expanded to address TB care comprehensively and is now named the Stop TB Strategy. The effectiveness of DOTS and the Stop TB Strategy is widely observed. The global target for TB is to cure 85 percent of TB patients who are given treatment.

Providing TB care is nonetheless a daunting task in countries such as Afghanistan and Somalia. Civil strife has seriously affected health infrastructure and consequently TB care. TB patients are often from the poorer segments of already impoverished communities. For them, accessing health and TB care is not easy. Treatment duration is also long: six to eight months. If patients default from treatment they may develop resistance to anti-TB drugs, particularly in a form of multi-drug resistance (MDR). MDR-TB is defined as resistance to, at least, the two most effective drugs, Isoniazid and Rifampicin; it is very difficult to treat. Patients may also develop resistance to second-line anti-TB drugs, those used for treating MDR-TB. This is called extensive drug resistance and is almost certainly untreatable.

Food-with-treatment for TB patients is an innovative way to address these challenges. In Afghanistan and Somalia, through a collaboration among WFP, WHO and national and local governments, patients receive food support during the course of treatment. This certainly improves the nutritional status of TB patients. Equally important, if not more so, food is an incentive for TB patients to continue the six to eight months of treatment.

Afghanistan

Afghanistan has managed to provide good care for TB patients and their families. In 2006, 25,475 TB patients were identified; 89 percent completed the eight-month treatment and were cured. This is an extremely encouraging result. Afghanistan has achieved the global target for TB treatment. Food support provided by WFP, WHO and the National TB Programme of the Ministry of Public Health is given to all diagnosed TB patients and extended to their families. This means a total of around 180,000 people suffering from TB directly and indirectly have received food support in Afghanistan in 2006.

Food is provided every one or two months during the course of treatment. The direct impact of food support on the high success rate (89 percent) is difficult to measure because all TB patients and their families receive food support in Afghanistan. It is not ethical to conduct a case-control study with two groups, one of which goes without food support. Still, anecdotal evidence on the impact of food support is abundant.

Somalia

In Somalia, TB care based on DOTS began in 1995. In collaboration with NGOs and local authorities, there is at least one TB centre in each of 18 regions in Somalia. In 2006, 11,945 cases of TB were diagnosed in these centres, 89 percent of whom completed the treatment and were cured.

Good TB care has been provided with the help of food support. The food support has contributed to the improvement of TB treatment in Afghanistan and Somalia, and it has been crucial for TB patients and their families in completing the long TB treatment. Even under extremely difficult circumstances, Afghanistan and Somalia have achieved the global target for treatment outcomes.

TB is often known as the disease of the poor, particularly in countries such as Afghanistan and Somalia. TB patients have to fight TB, but they also have to fight hunger. Hunger-related health problems are solvable when political will and action come together in the form of effective partnerships among national and local governments, NGOs, and United Nations agencies such as WHO and WFP.

Contributed to the World Hunger Series *by WHO Regional Office for the Eastern Mediterranean.*

"Whoever wishes to investigate medicine properly, should proceed thus: in the first place to consider the seasons of the year, and what effects each of them produces ... We must also consider the qualities of the waters ... and the mode in which the inhabitants live, and what are their pursuits, whether they are fond of drinking and eating to excess, and given to indolence, or are fond of exercise and labour ..."

Hippocrates, 460–380 BC

Linking childhood undernutrition with later life chronic disease

Chronic diseases are by far the leading cause of mortality in the world, accounting for 60 percent of all deaths (WHO, 2005). The primary causes of chronic diseases are well known: unhealthy diet, lack of physical activity and tobacco use. These causes are associated with raised blood pressure, raised glucose levels, abnormal blood lipids, overweight and obesity.

It is widely understood that a large proportion of the world's population is overnourished, with serious long-term detrimental health effects. Less well recognized is the relationship between an undernourished child today and later life incidence of obesity and chronic diseases such as heart disease, strokes, cancer, chronic respiratory diseases and type II diabetes.

Undernutrition and overnutrition increasingly coexist, especially in transition and middle-income countries. Referred to as the "nutrition transition", the double burden of early undernutrition and later overnutrition is particularly evident in countries undergoing rapid economic transitions, accompanied by alarming increases in chronic disease rates. The number of overweight people is rising in all parts of the world. In Latin America, India and China the relationship between undernutrition and overnutrition is of growing concern as their economies and societies rapidly transform (WHO, 2005).

Burden of chronic disease

- **35 million** people died from chronic disease in 2005 – half were under 70.
- **80 percent** of chronic disease deaths occur in low-income and middle-income countries.
- **1 billion** people worldwide are overweight.
- By 2010, **more obese people** will live in developing countries than in developed countries.
- Undernutrition, overnutrition and diet-related chronic diseases account for **more than half** of the world's diseases and hundreds of millions of dollars in public expenditure.

Sources: WHO, 2005, 2006a

This nutrition transition presents a unique challenge for countries that face a high prevalence of infectious disease and undernutrition, simultaneously with overweight and chronic disease. Global trends indicate that while obesity is predominantly a problem in high-income countries, its prevalence in urban areas in low-income and middle-income countries is now increasing among children and adolescents (Schneider, 2000). The costs associated with obesity-related diseases are high, especially for adults; however, prevention is essential during childhood and adolescence (Delisle et al., 2000).

Undernutrition and overnutrition converge, directly linked by hunger in the early stages of life.

There is increasing evidence to show that undernutrition during childhood is linked with overnutrition in later life. Extensive evidence from many countries also shows that conditions before birth and in early childhood influence health in adult life. For example, low birthweight is now known to be associated with increased rates of high blood pressure, heart disease, stroke and diabetes. There is evidence suggesting a connection between maternal undernutrition during pregnancy, rapid weight gain between 2 and 10 years of age, overweight in young children and an increased risk of chronic diseases in adulthood (Michels, 2003). Clearly, what transpires at one stage of the life cycle has implications for later stages.

Figure 19 – Malnutrition in children under 5 in LIFDCs

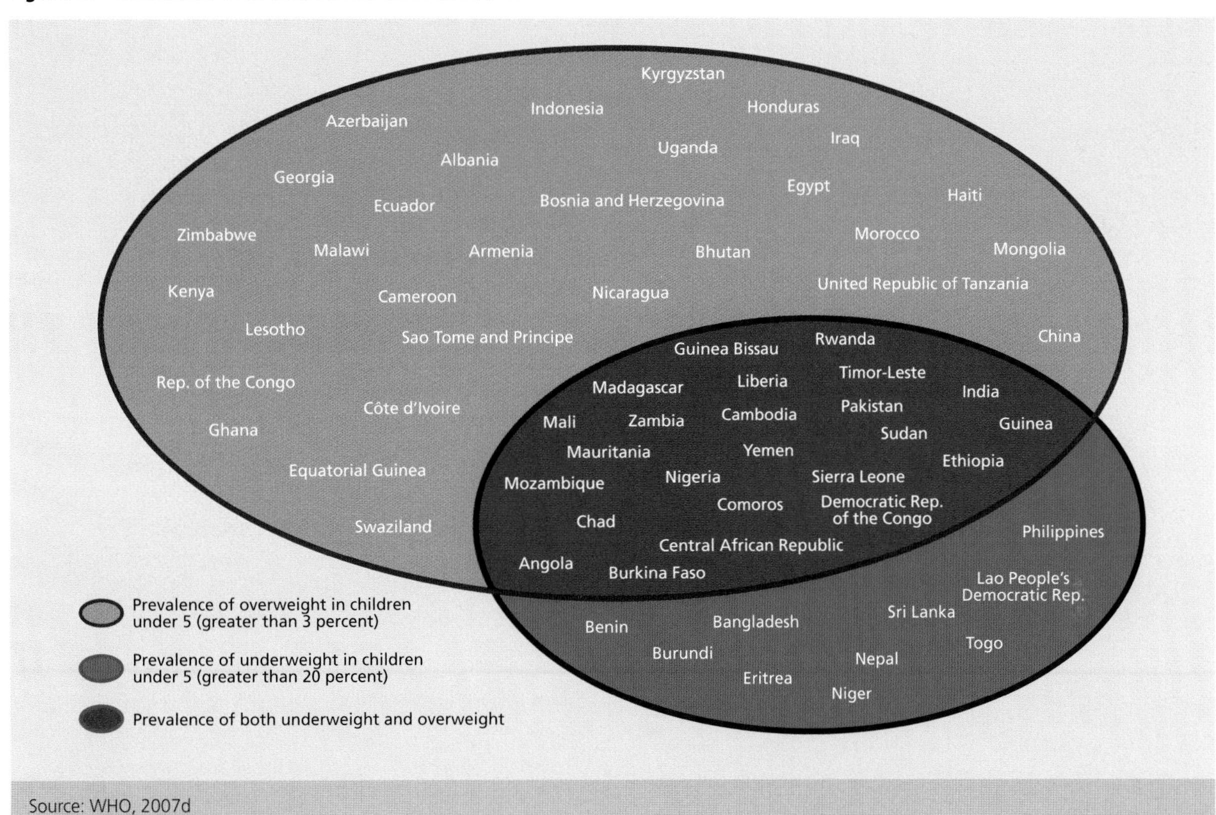

Source: WHO, 2007d

Early evidence that poor nutritional status in childhood is a predisposing factor for developing chronic disease later in life came from a number of studies during the 1944 Dutch famine. One study examined the effect of pregnant women exposed to food shortages who then developed glucose intolerance later in life (Godfrey and Barker, 2001). Another study looked at the relationship between coronary heart disease and birthweight (Roseboom et al., 2000a, b, 2001). These early findings, augmented by recent accumulated evidence, show a strong association between low birthweight and coronary heart disease, hypertension and type II diabetes.

A study in the United States of America found that girls aged 8–16 in food-insufficient households were 3.5 times more likely to be overweight than girls from food-sufficient households (Alaimo et al., 2001). This apparent contradiction is explained by the fact that cheaper foods often contain more energy, particularly in the form of fat and sugar, than more expensive foods of better quality. Obese children and adolescents are likely to remain so as adults, with a higher risk of disability and premature mortality (Sorenson et al., 1992; WHO, 2007c).

Obesity also has been associated with depression, especially among adolescent girls in the United States (Anderson et al., 2006).

A primary target group for obesity prevention should be adolescents, as health interventions in this age group appear to be more successful than in adult populations where lifestyle behaviours are set (Gortmaker et al., 1993; Delisle et al., 2000).

Knowledge gaps

Increased obesity among the hungry in developed countries may be associated with low physical activity and/or consumption of poor quality and high-fat diets, which tend to be cheaper and more accessible to low-income families. Further research is urgently needed to understand fully the relationship between hunger, adequate quality food, obesity and specific chronic diseases.

Intermezzo 6: Nutrition transition in Latin America – the experience of the Chilean National Nursery School Council Programme

Maria is worried that since she delivered Pedrito – her youngest son – she has been increasingly gaining weight. She also wishes she could be a couple of centimetres taller.

Maria's worry could be that of many women in Latin America. Over the past 20 years, the region has experienced several economic, demographic and environmental changes such as market globalization, urbanization and improved sanitation that have modified body composition, diet and physical activity patterns. High levels of undernutrition, energy-poor plant-based diets and intense physical activity are increasingly replaced by high rates of obesity and nutrition-related chronic diseases, high consumption of processed food and animals products, and sedentary activities – the process labelled the nutrition transition. Unfortunately, given the rapid speed of these changes, nutrition-related chronic diseases have compounded problems associated with nutritional deficiencies rather than replacing them, generating the double burden of disease. For example, the concurrence of adult obesity and stunting has been reported in Mexico and Brazil, and in most countries stunted children are overweight, rather than being under or normal weight for their height.

However, what is even more worrying is that early undernutrition seems to enhance the response to this nutrition transition. There is increasing evidence that exposure to undernutrition during critical periods of early life – the foetal stage and the first two years of independent life – predisposes individuals to obesity and obesity-related diseases in later life. Multiple studies have shown that thin babies and small and short infants have a higher risk of obesity, diabetes, cardiovascular disease and a number of other chronic diseases as adults. The challenge of addressing undernutrition involves simultaneously addressing problems related to undernutrition and overnutrition, and with a life cycle perspective.

Nutritional deficiencies and obesity are often rooted in poverty. Poor people in developing countries are still confronted with infections and undernutrition during childhood, yet if they make it into adulthood they will now be exposed to a lack of access to a healthy diet and few opportunities for engaging in physical activity, thus increasing their risk of obesity and obesity-related diseases. In some Latin American countries, for example Brazil, Chile and Mexico, obesity rates are already higher among low-income people, and economic projections suggest that this will eventually be the case in most countries in the region.

Yesterday, Maria was called to Pedrito's nursery school because he is overweight. His teacher explained to her the risks associated with childhood obesity and recommended that she decreases the amount of soda, chips and cookies that Pedrito eats at home. She also asked Maria to walk him to the nursery school. The teacher insisted on the importance of doing something now before Pedrito gets older and it becomes too late. Maria, however, is not sure: Pedrito looks like many other children at his nursery school ...

The Chilean National Nursery School Council Programme (JUNJI) is a national programme that provides childcare as well as supplementary feeding for low-income toddlers and pre-school children. In 1970, when the programme was founded, underweight and stunting were the most pressing nutritional problems for children in Chile, so the programme was oriented towards decreasing nutritional deficiencies. In the subsequent years – probably because of the contribution of many assistance programmes such as JUNJI – underweight in Chile declined from 16 percent in 1975 to 9 percent 1986; stunting – 5 percent in 1987 – remained a pending issue. As a consequence, in 1990 the programme moved away from using only weight-for-age indicators and incorporated length/height measurements into their periodic nutritional assessments.

In the next decade, the prevalence of stunting decreased while obesity increased. People in charge of JUNJI realized that nutritional problems among JUNJI children were no longer related to the "quantity" of food but to the "quality". Therefore, it was decided to decrease the number of calories provided by JUNJI meals. Reports from the JUNJI programme show that obesity rates stabilized after implementation of these changes. It is expected that additional adjustments to the programme such as workshops and training of parents may further contribute to decreasing obesity among this high-risk group.

Latin America has a network of nutrition assistance programmes that have successfully been tackling undernutrition for decades; examples are *Progresa* in Mexico and *Red de Protección Social* in Nicaragua. However, Latin America now faces the challenges associated with the nutrition transition. The JUNJI experience shows how adapting these ongoing nutrition assistance programmes and making use of already available resources can provide a unique opportunity to respond to the challenges. As countries move into further stages of the nutrition transition, they should focus on adapting new strategies to ensure adequate nutrient content of the food – energy and micronutrients – rather than cutting nutrition assistance programmes.

Data that are routinely collected as part of these programmes should be used to monitor trends and the actual nutritional status of the beneficiaries in order to re-define their objectives and priorities. This will be the only way to continue to ensure adequate nutrition for many Marias and Pedritos.

Contributed to the World Hunger Series *by Dr Camila Corvalan, Emory University, University of Chile.*

"In order to reduce hunger and poverty and increase access to clean water and sanitation, we need to have a strong base of environmental sustainability which is providing these services on which people rely for their well-being."

Neville Ash, 2005

Global climate change, avian influenza and urbanization are among the current transitions that have the potential to dramatically alter efforts to eliminate hunger and poor health. As with previous transitions and threats, it will be the hungry and marginalized who will be most affected.

Global climate change

Over the past two decades, developed and developing countries have experienced the effects of climate change in the increasingly frequent occurrence of high-impact natural disasters (Guha-Sapir et al., 2004). In addition to increased frequency and severity of natural events, other worrying threats that also have the potential to thwart hunger reduction strategies are the decreasing availability of clean water and arable land, and reduced biodiversity.

Alterations in climate will bring about changes in the amount of rainfall and the availability of water from glaciers and snow. Water availability and quality are projected to decrease, particularly in regions affected by drought. By 2020, between 75 million and 250 million people across Africa could face water shortages (IPCC, 2007). Changing climatic conditions, increasing natural disasters, and over-exploitation of the world's biodiversity may challenge the capacity of ecosystems to adapt.

Decreasing biodiversity

Biodiversity and human well-being are inextricably linked. Just as our ancestors relied on a variety of species, the modern world depends on and benefits from ecosystem services. Before stable settlements were established, early humans could select from about 250 plant species for a diversified diet that contributed positively to their health. However, by the 3rd millennium BC the number of available plant species consumed had dwindled to 56. Globally, approximately 75 percent of our calories today come from no more than 12 plants (Barnes, 2007).

In some areas, maintaining dietary diversity may become the pressing challenge. "Roughly 20–30 percent of species are likely to be at high risk of *irreversible extinction* if the global average temperature rises by 1.5–2.5°C beyond 1990 levels. For increases in global average temperature exceeding 1.5–2.5°C, there are very likely to be major changes in ecosystems which will adversely affect the environmental goods and services which humans use" (IPCC, 2007).

Affecting food availability

Global and local food production will be affected by rising temperatures and changes in the amount of rainfall and water availability. Crop yields in higher latitudes will probably increase if temperatures rise by 1–3°C. Conversely, yields are expected to decrease in lower latitudes, particularly in the seasonally dry tropics (IPCC, 2007). Combined with the potential for an increase in the number of natural disasters, and the damage they may cause to agriculture production, changing agricultural patterns may lead to increased food shortages and hunger (Pimentel, 1993).

Ecosystem changes brought on by climate change may also alter the prevalence of pest infestations, negatively affecting food production. Climate change may modify soil temperatures and moisture levels and alter the success of beneficial organisms and pests. A rise in temperature will prolong breeding seasons and reproductive rates, increasing the number of insects. Climate change could extend the range of insects and pests into new ecosystems. This could have consequences for the spread of infectious diseases and for agriculture.

Another threat to food availability

Concern over high petroleum prices and unstable supply, geopolitical tensions and growing environmental awareness have increased interest in bio-fuel production. Although the growth of a bio-

Losses and food security

The destruction of agricultural products brought on by insects can be devastating. For example, in 2004 locusts invaded West Africa and devastated crops, fruit trees and vegetation (FAO, 2006). Developed countries are not immune to agricultural losses associated with pests; a study estimates that farmers in the United States will experience a 25 to 100 percent increase in losses depending on the crop (Pimentel, 1993).

energy industry is good news for those who grow the crops, including farmers in developing countries, higher food prices and the diversion of food crops for energy purposes could reduce the overall availability of food. Further, it could undermine poor people's access to quality food and worsen the economic situation for low-income net food-importing countries. In fact, many small farmers in developing countries are net buyers of food and could be negatively affected by higher food prices.

In developing countries, bio-fuel production may provide an incentive for substantial shifts in crop production, largely at the expense of food and animal feed. Bio-fuel production could also have a major impact on land use and those dependent on agricultural economies.

Food assistance shipments are influenced by crude oil prices and grain prices. If donors' food assistance budgets are fixed in value terms, higher energy and food prices could mean smaller quantities of food assistance, which is likely to have an impact – at least in the short term – on food procurement and resourcing. The result may be decreased availability of food assistance, and consequently increased food insecurity for the most vulnerable and hungry (WFP, 2007).

Exposure to new diseases

Climate change may result in warmer temperatures at higher altitudes and latitudes, which in turn could alter the average exposure of human populations to disease. For example, climatic patterns are closely linked to the life cycle of the malaria parasite: the macro-environment determines the type of malaria transmission that occurs, which affects a community's vulnerability to malaria infection. Countries on the borders of endemic zones, such as desert fringes and upper highland limits, may become prone to malaria epidemics with increased frequency (Bates et al., 2004).

Figure 20 – Historic relationship between grain prices and food aid volumes

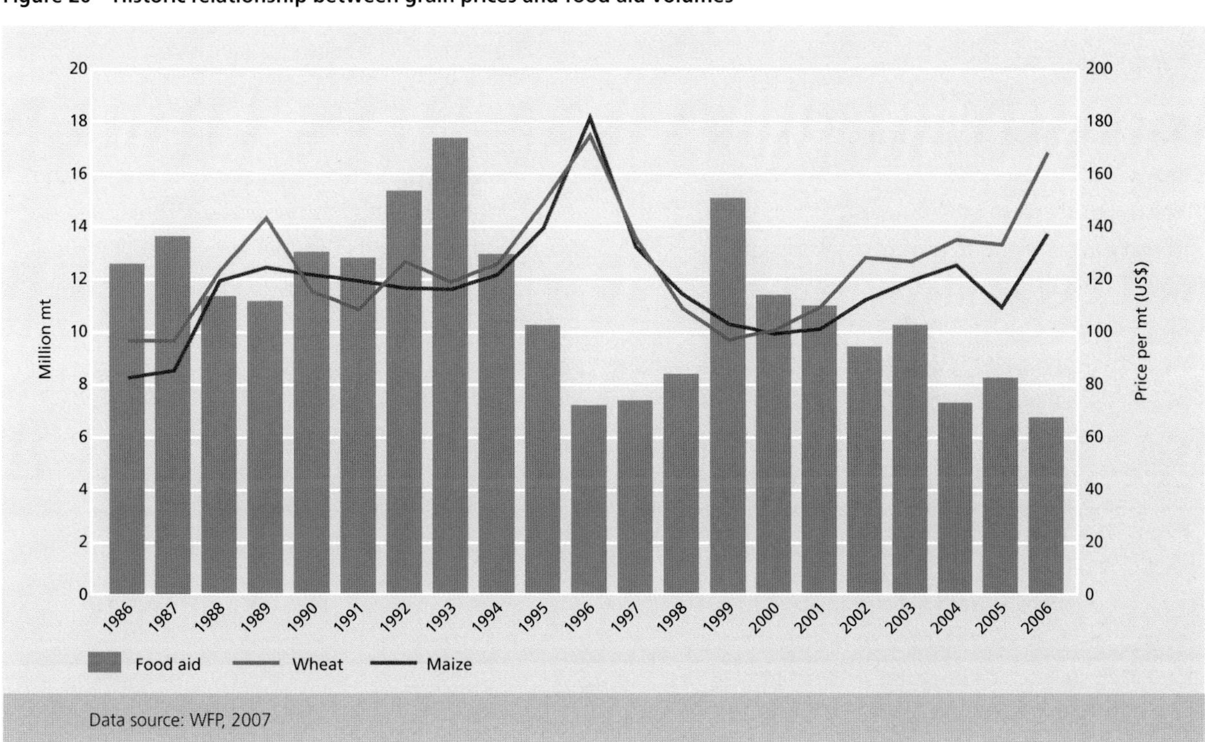

Data source: WFP, 2007

Avian influenza

Despite control measures, avian influenza continues to spread in birds, raising serious public health and food security concerns over unsanitary coexistence of people and animals (FAO and OIE, 2005).

Poor health infrastructure in many developing countries, compounded by undernutrition, is unlikely to cope adequately with emerging and re-emerging disease, and in some cases may make the re-emergence of disease more prevalent.

Human transmission would create enormous new challenges for health systems and services, particularly in Africa, where health systems are already facing the demands of AIDS, tuberculosis and malaria (WHO, 2006b).

Food insecurity and poverty can increase the likelihood of unsafe food practices related to poultry production and consumption. In poor households, poultry and eggs are inexpensive sources of protein and a means of generating income that is not capital-intensive; millions of people live alongside chickens, increasing the chances of the virus crossing over to humans (FAO and OIE, 2005). An outbreak of avian influenza in poultry would devastate livelihoods and increase hunger.

HUMAN CASES OF AVIAN INFLUENZA

Country	Cases	Deaths
Azerbaijan	8	5
Cambodia	7	7
China	25	16
Djibouti	1	0
Egypt	36	15
Indonesia	100	80
Iraq	3	2
Lao People's Democratic Rep.	2	2
Nigeria	1	1
Thailand	25	17
Turkey	12	4
Viet Nam	93	42
Total	**313**	**191**

Cumulative number of confirmed human cases from 2003 to 2007, as of 15 June 2007.
Source: WHO, 2007e

Urbanization and slums

The year 2007 marks an historic transition: the number of people living in urban communities exceeds the number living in rural areas. More than 3 billion people now live in urban agglomerations (UN-HABITAT, 2006). Asia, Latin America and Africa have experienced rapid and unplanned development of so-called "mega-cities" of more than 20 million people. Trend projection suggests that mega-cities and middle-sized cities in the developing world will absorb 95 percent of urban growth in the next two decades; by 2030, such cities will be home to 80 percent of the world's urban population.

Forced displacement, increasing populations, rapid economic development and rural-to-urban migration play a critical role in urban development and the growth of urban poverty and slum formation. The relationship between hunger and poor health becomes a stark reality for those with no means to produce food, no employment opportunities and no access to quality healthcare. Poor sanitation, undernutrition, crowding and lack of adequate healthcare facilities coalesce to create an environment ripe for common infectious diseases that further threaten nutritional status.

Children living in slums face risks from water-related and respiratory illnesses. In Ethiopia, undernutrition in children under 5 in slums is 47 percent, compared with 27 percent in non-slum urban areas. In Brazil and Côte d'Ivoire, the prevalence of child undernutrition is three to four times higher in slums than in non-slum areas (UN-HABITAT, 2006).

Livelihood options also appear to be constrained by where people live. One study in France showed that job applicants residing in poor neighbourhoods were less likely to be called for interviews than those who lived in wealthier neighbourhoods. Another study in Rio de Janeiro found that living in a slum was a greater barrier to gaining employment than being dark skinned or a woman – a finding that confirms that "where we live matters" when it involves access to food, basic healthcare, education and employment (UN-HABITAT, 2006).

Challenges ahead

Part II has set out the close relationships of hunger and health, and undernutrition and disease, drawing on evidence related to some of the threats facing the world today. Throughout history, humans have advanced and confronted new transitions. Today, the combined impact of undernutrition, overnutrition and disease, alongside global warming and urbanization demand that leaders confront these problems head on – and soon.

Figure 21 – Number of slum dwellers (actual and projected, 1990–2020)

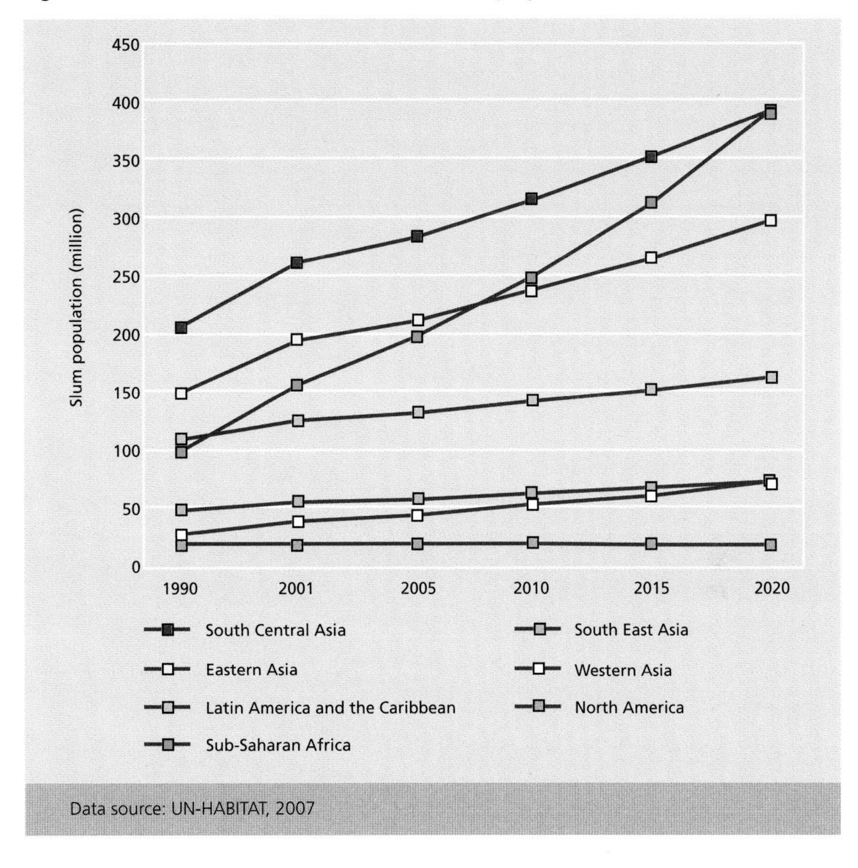

Data source: UN-HABITAT, 2007

Part III National Development: Commitment and Political Choice

The challenge, therefore, is for all those involved in nutrition and health to optimize resources to eliminate hunger for all people …

Part III furthers the discussion of the consequences of hunger and poor health by delineating their impact on national development. **Chapter 1** outlines the debilitating economic costs of hunger, in particular the human capital loss, and the brake this puts on national development. **Chapter 2** highlights proven cost-effective activities – "essential solutions" – for addressing the interrelated problems of hunger and poor health. **Chapter 3** describes some of the choices before leaders, and commitments which, if taken seriously, could make significant strides towards eliminating hunger.

3.1 Hunger impacts on human development

"The problems we face today, violent conflicts, destruction of nature, poverty, hunger and so on, are human-created problems which can be resolved through human effort, understanding and the development of a sense of brotherhood and sisterhood. We need to cultivate a universal responsibility for one another and the planet we share."

14th Dalai Lama (b. 1935)

The high cost of hunger

Hunger and poor health, undernutrition and disease weave layer upon layer of burden on the poor – affecting individuals, families, communities, and ultimately national development and growth – resulting in massive human capital loss, generation after generation. The high costs associated with the treatment of disease and undernutrition are serious constraints to development.

For the first time in history, the world can direct enormous resources to overcoming hunger and poor health. There is growing recognition that the cost of inaction is high, in economic and in moral terms – and that the cost of action is modest by comparison. A number of proven solutions are available and affordable, but they have to be scaled up to reach the world's vulnerable and marginalized people. An enabling environment to convert knowledge into feasible action and to remove institutional blockages is essential; otherwise, it will be difficult to maximize the potential gains from growing public and private resources to tackle hunger and poor health. Synergistic approaches are needed to make the best use of the increased resources in implementing proven solutions. And leaders need to make the right political choices.

Hunger and productivity losses are closely related

Hunger and poor health directly affect human and social capital formation and economic growth. As the *World Hunger Series 2006: Hunger and Learning* showed, undernourished children may receive less schooling and thus earn less during their lifetimes. The effects are long-lasting and inter-generational, with impacts impeding the achievement of other global social goals (Fernholz et al., 2007). Reducing hunger increases productivity by improving learning, cognitive development and work capacity, and reduces the impact of disease and premature mortality.

Underweight is the largest risk factor contributing to the burden of disease in developing countries. A number of studies have documented the high economic cost of hunger and poor health and the indisputable link with economic productivity and output (Hall and Jones, 1998; Behrman et al., 2004; Edwards, 1998; Barro, 1990). For example, there is a 17 percent loss in manual labour productivity as a result of iron deficiency, as shown in Figure 22.

The strongest evidence for the hunger–productivity relationship is related to growth in early life. Stature, or height, has been shown to be related to productivity: a 1 percent reduction in adult height as a result of stunting is associated with a 1.4 percent loss in productivity (Bloom et al., 2001).

Undernutrition has also been implicated in losses of more than 10 percent of individual lifetime earnings as a result of impaired physical and cognitive productivity (World Bank, 2006). An example is anaemia, which has been associated with lower productivity even in tasks requiring moderate effort. In South Asia, for instance, productivity losses due to iron deficiency alone are associated with an estimated loss of about US$4.2 billion annually (Horton and Ross, 2003). These figures can be translated into losses of up to 3 percent of gross domestic product (GDP), depending on the size of the economy concerned.

Height linked to earnings

Adult height as a measure of stunting has been linked to level of earnings, broadly showing that for men and women who work in the market sector (in urban Brazil), a 1 percent increase in height leads to a 2.0–2.4 percent increase in earnings (Thomas and Strauss, 1997).

Figure 22 – Undernutrition and lifetime loss in individual productivity

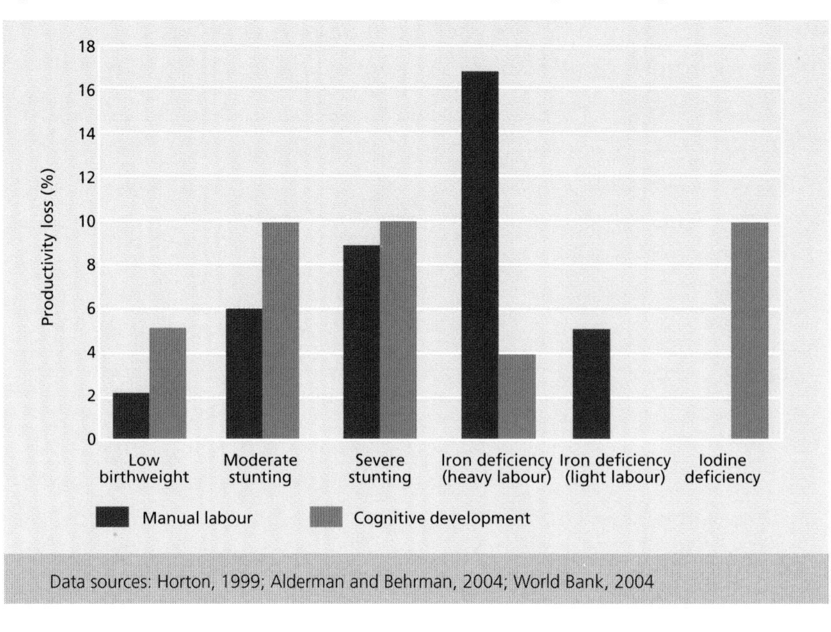

Data sources: Horton, 1999; Alderman and Behrman, 2004; World Bank, 2004

An analysis in India estimated that the cost of productivity losses, as a result of disease and death related to malnutrition, were between US$10 billion and US$28 billion a year, or 1.0–3.6 percent of GDP (Chatterjee and Measham, 1999).

The assessment, while informative, cannot account for all the multi-dimensional impacts of hunger and malnutrition, present and future. Consequently it probably underestimates the consequences of inaction and the benefits of action.

In a study of 12 countries, household-level data were modelled to predict the impact of GDP growth per capita on the prevalence of undernutrition in pre-school children over a 20-year period (Alderman et al., 2001). It showed that even with sustained optimistic GDP growth per capita from 2000 to 2020, and excluding all other factors at current rates of progress, the declines in undernutrition would not be sufficient to reach the related MDGs.

These analyses highlight that economic growth is necessary but not sufficient. Inertia will outpace hunger reduction if economic growth remains the dominant strategy.

Improved nutrition also generates savings

The economic benefits of actions to reduce hunger and improve nutrition can also be viewed as savings. Countries gain because of increased productivity and because they do not have to incur certain costs such as healthcare costs. Undernutrition in children and adults can generate healthcare costs through increased frequency of medical treatment and hospitalization. But the implications go beyond productivity and healthcare costs: hunger and poor health leave people unable to cope with external shocks such as natural disasters.

In collaboration with LAC governments, WFP and the Economic Commission for Latin America and the Caribbean (ECLAC) jointly developed a methodology to estimate the economic impact of child undernutrition for a given year. Based on national data, losses amounted to over US$6.6 billion in 2004 for the seven countries included in the study (ECLAC and WFP, 2007).

Productivity losses account for 93 percent of the total cost. These losses are spread almost equally between higher mortality rates and lower educational levels. Costs caused by higher mortality rates resulted from 2.6 million cases of premature death attributable to causes related to undernutrition, of which it was

ESTIMATED COST OF CHILD UNDERNUTRITION IN LAC, 2004

	Costa Rica	Dominican Republic	Guatemala	Honduras	Nicaragua	Panama	El Salvador	Total
US$ million	318	672	3128	780	264	321	1,175	6,659
% of GDP	1.7	3.6	11.4	10.6	5.8	2.3	7.4	6.4

Source: ECLAC and WFP, 2007

ESTIMATED SAVINGS FROM REDUCING CHILD UNDERNUTRITION IN LAC, 2004–2015 (US$ MILLION)

	Costa Rica	Dominican Republic	Guatemala	Honduras	Nicaragua	Panama	El Salvador	Total
Achieving MDGs	49	71	525	118	25	99	133	1,019
Eradication	49	71	1,534	243	46	125	203	2,271

Source: ECLAC and WFP, 2007

assumed that 1.7 million cases were among the 2004 working-age population. These represent a 6 percent loss in the working population in the sub-region for that year. Deficiencies attributable to lower educational levels were calculated on the basis of an average of two years less schooling completed by undernourished people.

The method also allowed for estimates of different future scenarios. These show that attaining the undernutrition target of MDG 1 would create savings of US$1.02 billion, and that eradication of child undernutrition would yield savings of over US$2.27 billion between 2004 and 2015. If the timeline is extended beyond 2015, the potential savings grow even larger.

This huge economic toll combined with the enormous savings generated from actions to address hunger provides a strong impetus for strengthening alliances among governments, private sector and civil society to initiate key actions to reduce the devastating impacts of hunger.

3.2 Effective solutions

"The doctor of the future will no longer treat the human frame with drugs, but rather will cure and prevent disease with nutrition."

Thomas Edison (1847–1931)

Solutions throughout the life cycle

All too often, projects are designed around one main activity and are unable to bring together the benefits of complementary activities. The *World Hunger Series 2007* sets out proven, practical and cost-effective solutions to address the interrelated causes of hunger and poor health. These solutions, combining food-based activities with basic healthcare and prevention, form "essential solutions" for the reduction of hunger and poor health. Conceptually similar to the "essential package" in support of school feeding programmes developed by the United Nations Children's Fund (UNICEF) and WFP, these essential solutions are known and are effective individually – but combined in a package, they can generate substantial additional benefits.

With an emphasis on impact throughout the life cycle, these essential solutions are intended to prevent hunger and improve the health of hungry people and contribute to achieving the MDGs. The intention is to expand programmes aligned with two broad "windows of opportunity" – critical times in an individual's life: ***early life***, focusing on mothers, infants and young children, and ***adolescence***, which includes school-age children. However, effective solutions are proposed for all stages of the life cycle.

The main benefits of these essential solutions are that they increase the effectiveness of other investments in human capital, lead to better health outcomes and improve social equity. The costs are indicative, based on actual implementation figures in a given location. In poor countries, the resources required to eliminate severe forms of hunger and nutritional deficiencies may be commensurably larger and more difficult to secure. This is probably because the cost of basic health services varies greatly depending on country, region, proximity to transport and supply routes, geography, topography and population density. Thus it is difficult to present absolute cost figures; average costs have huge deviations and are often higher when programmes reach the most in need. Nevertheless, the

Figure 23 – Practical solutions for all stages of the life cycle

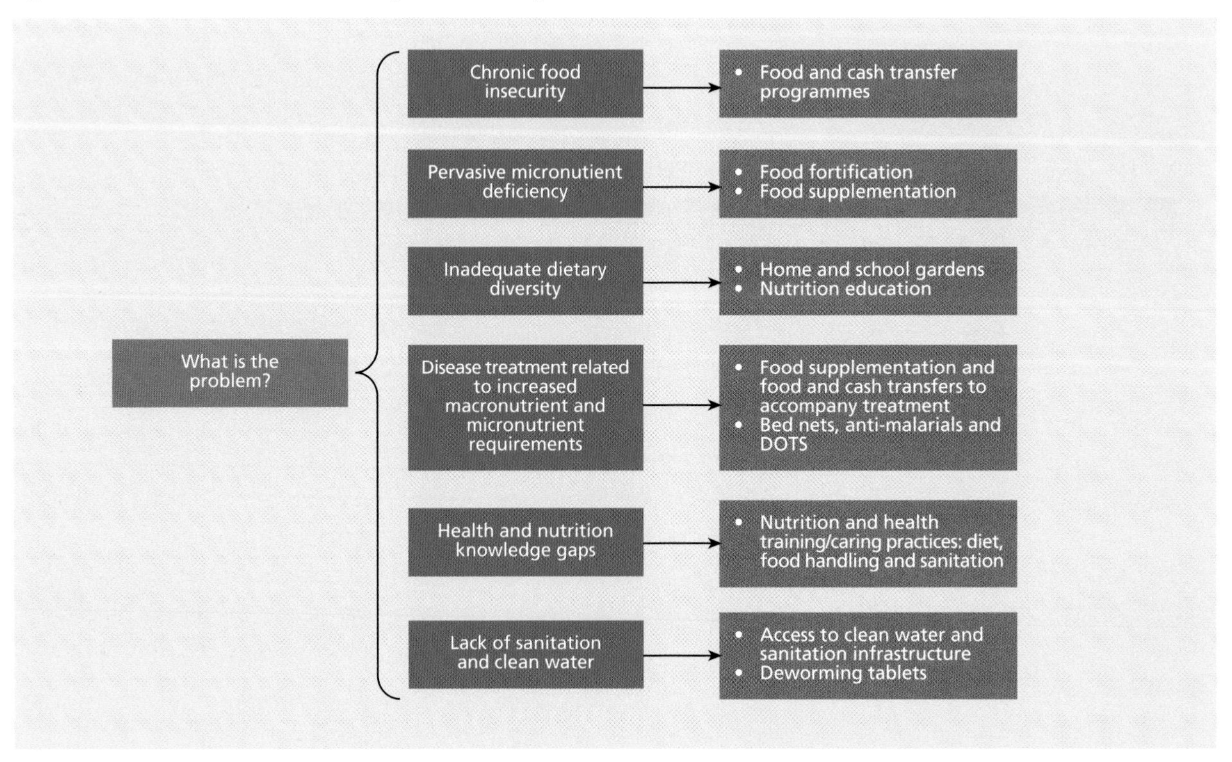

Essential solution	Indicative costs
Food and cash transfers	Ration for person entirely reliant on food assistance: US$5.10 per person/month (WFP, 2000, 2006).
Targeted micronutrient supplementation	Iron: US$3.17 to US$5.30 per child/treatment (Horton, 2006).
	Iron + folic acid: US$0.027 per tablet (Gillespie et al., 2007).
	Vitamin A: US$0.25 to US$0.67 per child/dose (World Bank, 2004).
	Zinc: US$0.47 per child/treatment (Gillespie et al., 2007).
Supplementary feeding	On-site ration US$2.09 per person/month (WFP, 2002, 2006).
	Take-home ration US$3.11 per person/month (WFP, 2002, 2006).
Complementary feeding	US$3.11 per child/month (WFP, 2002, 2006).
Large-scale food fortification	Iodine: US$0.10 per person/year (WHO and FAO, 2006).
	Iron: US$0.12 per person/year (WHO and FAO, 2006).
	Vitamin A: US$0.10 per person/year (WHO and FAO, 2006).
	Zinc (wheat flour): US$0.06 per person/year (WHO and FAO, 2006).
Home fortification	Supplementation packet: US$0.015 to US$0.035 per packet (Zlotkin, 2007).
Nutritional support for disease treatment (HIV/AIDS and TB)	Nutritional support for ART: US$4.99 per child/month (WFP, 2006).
	Nutritional support for TB: US$4.99 per child/month (WFP, 2006).
Infectious disease prevention:	
Malaria	Anti-malaria drugs: US$0.10 to US$0.20 per treatment (Gillespie et al., 2007).
	Bed nets: US$2.50 to US$3.50 per bed net (Gillespie et al., 2007).
Tuberculosis	DOTS: US$128 per treatment (Gillespie et al., 2007).
Parasites	Deworming tablets and delivery: US$1.40 per child/year (WFP, 2007).
Diarrhoeal disease	ORT: US$0.08 to US$0.10 per treatment (Gillespie et al., 2007).
HIV/AIDS	Male condoms: US$0.03 per unit (UNFPA, 2005).
Diversify diets and promote quality food consumption	School meal: US$20 per child/year (WFP, 2007).
	School gardens – as part of full essential package: US$16 per child/year (WFP, 2007).
Transfer of knowledge on healthcare and food practices	Breastfeeding promotion: US$2 to US$3 per child (Caulfield et al., 2006).
	Community-based nutrition programmes: US$2 to US$10 per child (Caulfield et al., 2006).
	Deworming training for teachers: US$0.78 to US$1.08 per capita per year (Disease Control Priorities Project, 2006).
	Education on HIV: US$0.48 per child/year (WFP, 2006).
	Nutrition education for pregnant women: US$3.75 per woman/year (World Bank, 1994).
Access to clean water and improved sanitation	Point-of-use water treatment: US$170 to US$525 per unit (EPA, 2007).
	Simple pit latrine: US$415 per unit (EECCA Working Group, 2006).

Sources and calculation methods: see pages 199–201

solutions prescribed in the table above and discussed in this chapter are known, effective and affordable.

Solutions for all stages of the life cycle

The proposed essential solutions are practical and appropriate for all stages of the life cycle. It should be noted that they combine food-based solutions with those that address broader health problems. Access to basic healthcare and addressing hidden hunger with micronutrient supplementation are fundamental features.

Practical solutions that have benefits throughout the life cycle include providing bed nets treated with insecticide to combat malaria and other vector-borne diseases, diversifying diets for more balanced nutrition, enabling access to clean water and increasing knowledge of sanitary practices. Deworming tablets also should be made readily available. Health and nutrition training, including support for food and childcare practices related to diet, food and a healthy lifestyle, is a critical complement to basic healthcare at all stages of life.

Food and cash transfers. Targeted food and cash transfers help vulnerable households to cope with shocks and meet their minimum food consumption requirements. Support can be provided through targeted general food distributions, blanket supplementary feeding, mother-and-child health and nutrition programmes, institutional feeding, food for work and cash for work, food or cash for training, and school feeding.

Gender sensitivity is paramount in food-based and cash-based transfer programmes. Distribution to women, for example, is recommended for two main reasons: first, women often have the primary role in household food management; and, second, women are more likely to use food rations for nutritional purposes as opposed to sale or exchange for other goods. When food and cash transfers are provided, certain measures can improve their effectiveness, specifically ensuring that:

- transfers are not associated with risk of attack or abuse;
- special arrangements are made for pregnant women, women with young children and the elderly;
- women are able to transport the food home – distances to distribution points must be short and food packages must not be too heavy; and
- the programme does not interfere with women's other domestic responsibilities (WFP, 2000).

Supplementary feeding. Targeted supplementary feeding programmes are directed at selected individuals at risk, aiming to:

- rehabilitate moderately undernourished people, particularly children and adolescents;
- prevent moderately undernourished people from becoming severely undernourished;
- reduce the risk of morbidity in children under 5;
- provide a food supplement for pregnant and lactating women and other individuals at risk; and
- provide follow-up to referrals from curative feeding programmes.

Fortified maize meal

A 2003 pilot project in Zambia showed that fortifying maize meal using mobile mills in a refugee camp greatly improved health and nutritional status. Among children there were improvements in height and weight, anaemia was reduced from 47.7 percent to 24.3 percent and vitamin A deficiency was reduced from 46.4 percent to 20.3 percent. Rates of illness among women and children decreased, and pregnancy outcomes improved (WFP, 2007).

For example, a child under 3 needs to obtain 25–35 percent of energy from fat. All children need foods rich in vitamin A: typically, breast milk, animal source foods (eggs, liver, whole fish, dairy products), green leafy vegetables and orange-coloured fruits and vegetables. After an illness, a child needs extra meals every day for at least a week. Supplementary feeding is more cost-effective when integrated with actions that cover the non-food related causes of undernutrition.

Food fortification. Food fortification is the process whereby one or more micronutrients are added to commonly eaten foods to improve the quality of the diet. Fortification increases micronutrient intake, to compensate for a low intake or bioavailability of one or more nutrients. Requirements may be high because of growth or infections.

A suitable food vehicle is one that is acceptable, affordable and frequently consumed by the population and that can be made available through an effective distribution system. Unprocessed foods such as whole grains or pulses may be difficult to fortify, especially at the sub-national level; but sugar is frequently fortified with vitamin A at the national level in Nigeria and Zambia and in some Central American countries. For example, WFP fortifies processed foods such as salt, oil, cereal flour, blended foods and biscuits, and is increasingly engaged in local production of fortified blended foods. Trials are ongoing for the fortification of rice.

Fortification is not the only option for providing missing nutrients: providing supplements may be a more effective strategy, depending on the type of deficiency and the local circumstances.

Addressing chronic food insecurity through productive safety nets

The prevalence of undernutrition remains very high in the north-eastern highlands of Ethiopia, where 52 percent of children under 5 are reported to be underweight, and in the Amhara region, where 57 percent are reported to be stunted. Since 2005, Save the Children has been one of the implementing partners for the national Productive Safety Nets Programme. It is one of the government's priority programmes to tackle hunger; its central component provides cash or food in exchange for work with the objectives of providing transfers to chronically food-insecure populations (so that they can meet immediate needs and prevent asset depletion), and creating assets for the community.

Complementary interventions enhance the impact of the programme in terms of children's nutritional status and included livelihood-promotion activities such as providing sheep or goats, chickens and bees, and nutrition education for mothers to improve childcare.

Meket *Woreda* of Amhara National Regional State (2004–2006)

The cash-for-work project in Meket *Woreda* (administrative district) made cash payments of 1,106 birr (about US$125) per household to 70,000 vulnerable and chronically food-insecure people. Detailed monthly expenditure data collected from 50 households showed that almost 76 percent of family budgets were spent on foods such as sugar, oil, potatoes and meat. Mothers indicated that as a result of the cash intervention they were able to feed their children from a wider food basket that included a greater variety of grains and pulses, animal products and oil, and increase the frequency of meals. Families could also purchase soap and clothes and improve their access to medical care.

In addition to the standard cash-for-work programme, a complementary component to protect and promote breastfeeding was implemented. An analysis of the causes of undernutrition before the project began found that for children under 6 months, breastfeeding was the most important determinant of undernutrition: children who were not exclusively breastfed were five times more likely to be undernourished. Poorer women reported that they had to leave their homes earlier after giving birth and for longer periods to obtain food and income for their families, and were less able to breastfeed exclusively for the first six months of their children's lives. In response, the programme provided an intervention whereby women who were breastfeeding could be relieved from working for cash and instead could attend nutrition and education sessions with their babies, thereby allowing them to breastfeed for longer.

Contributed to the World Hunger Series *by Save the Children UK.*

Food supplementation. A food supplement is typically a nutrient added to a food that does not contain it. Several types of food supplements are recognized: (i) additives that return a deficit to "normal" levels, (ii) additives that enhance the nutrient value of a food, and (iii) supplements taken in addition to the normal diet.

Along with fortification, supplementation is a vital component in addressing the nutrient needs of critical target groups. Sustainable and effective fortification and supplementation programmes include (Sanghvi et al., 2007):

- intervention packages built around the two proven core intervention approaches – fortification and supplementation – recognizing that people obtain micronutrients through multiple channels; and
- community approaches to supplementation and strategies such as intensified outreach and social mobilization to assure coverage of marginalized populations.

Despite the success of fortification, new strategies are needed to meet the varied micronutrient needs of vulnerable populations. Groups such as children under 2 have greater micronutrient needs that sometimes are difficult to meet with current strategies. New products are being tested to respond more effectively to micronutrient deficiencies: in particular, new formulations of blended foods are being revised and developed.

Home fortification is a promising strategy for delivering micronutrients to children whereby micronutrient powders and spreads are added to food prepared at home. Experiences from Darfur and Indonesia show that food can be fortified for an individual or the whole family with a precise amount of micronutrients.

Diversifying diets. Diversifying diets can help to prevent micronutrient deficiencies. The promotion of school and home gardens, agricultural skills training and nutrition education are cost-effective solutions that help to increase diversity in diets. An essential element is increasing the consumption of vegetables and fruits, poultry, fish and small animals, thus ensuring a greater variety of vitamins and minerals that are not normally found in the staple diets of hungry people. Other ways of diversifying diets include:

- augmenting nutrients through plant breeding;
- improving diversity in agriculture production through home and school gardens; and
- training in home food processing and storage.

Nutritional support with disease treatment. Disease treatments such as ART and DOTS should be accompanied with nutritional support. Even without treatment, those affected have higher energy and micronutrient requirements. As discussed in Part II, evidence indicates that ART and DOTS require a continued high nutritional intake to be effective and sustainable. Thus adequate quality food is important to optimize the benefits of these treatments.

Many lives can be saved with adequate nutrition

By the time patients become eligible for treatment, weeks or months of sickness have exhausted their financial resources, making it difficult to find the food to consume with the drugs. In Kenya and Mozambique, for example, some people do not want to begin treatment because they do not have enough food to support the increased appetite that comes with ART and DOTS (WFP, 2006a).

Treated bed nets and anti-malarials. Insecticide-treated bed nets are an inexpensive way to prevent people from contracting malaria. They have been demonstrated to reduce significantly malaria morbidity and mortality, and thereby to improve nutritional status. Impregnated bed nets, for example, have been shown to reduce significantly maternal and placental malaria and maternal anaemia, resulting in reduced risk of low birthweight in newborns (Ter Kuile et al., 2003).

Expanding the use of bed nets

Population Services International developed a novel approach to increase the use of insecticide-treated bed nets by the poor in Malawi. The programme sells bed nets for US$0.50 to mothers through rural antenatal clinics; the nurse who distributes the nets receives US$0.09 per net, providing an incentive to ensure that nets are always in stock. Nets are also sold to richer Malawians through private-sector channels at a higher cost (approximately US$5.00); these profits are used to subsidize the nets sold through the antenatal clinics, so the programme is self-funding. The programme helped to increase the nationwide average of children under 5 sleeping under nets from 8 percent in 2000 to 55 percent in 2004, with similar increases for pregnant women. A follow-up survey found nearly universal use of the nets by those who had paid for them. By contrast, a study of a programme to hand out free nets to people in Zambia found that 40 percent of the recipients did not use the nets (Easterly, 2006).

Insecticide-treated bed nets do not constitute a stand-alone solution for malaria, but their remarkable impact in terms of reducing malaria transmission and saving lives suggests that they should be a basic part of the health programme in every malaria-endemic area (Hawley et al., 2003).

Deworming tablets. Deworming treatments, usually in tablet form, are an effective and inexpensive way to address micronutrient deficiencies, including anaemia,

Examples from Uganda and Afghanistan

An analysis of the cost-effectiveness of a nationwide school-based deworming programme in Uganda showed that the cost per child was between US$0.32 to US$0.70 per treatment, very little compared with the health benefit to these children (Brooker et al., 2007).

The national deworming campaign, a joint initiative of the Government, WFP, UNICEF and WHO, carried out a health and hygiene education training session for 8,800 teachers; subsequently, 6 million children received deworming treatment in 2006 (WFP, 2006b).

resulting from poor nutrient absorption caused by intestinal worms. Chronic infestation can lead to long-term retardation of mental and physical development. Severe worm infestation can lead to death (UNICEF, 2007c). An effective way of providing deworming treatment is through school programmes.

Clean water and improved sanitary practices. Unsafe water and inadequate sanitation, which affect over 1.1 billion people, are basic problems that are directly linked to the causes of undernutrition – the quality and quantity of food consumed and the transfer of disease. Children bear the brunt of the burden – the majority of people who contract water-related disease are under the age of 5, with the highest proportion under the age of 2 (Disease Control Priorities Project, 2007).

Evidence from Kenya

A safe water system was introduced in Kenya to reduce the risk of diarrhoea. Teachers taught students about safe water and hygiene practices. Water-storage vessels were placed between classrooms and filled daily with treated water. Clinic visits for diarrhoea peaked between January and March 2002 at 130 visits and in 2003 at 71; but in 2004, after project implementation, only 13 diarrhoeal episodes were recorded (Migele et al., 2007).

Infections transmitted by water are the leading cause of diarrhoea. When combined with reduced access to preventive and curative healthcare, the impact of unsafe water and inadequate sanitation is profound. Improving access to clean water and promoting basic hygiene and sanitation are essential solutions for reducing hunger and improving health.

Windows of opportunity

Early life: solutions for mothers, infants and young children

The first window of opportunity – early life, which includes pregnancy, infancy and early childhood – is apparent from recent demographic and health surveys (DHS) from Bangladesh, Cambodia, Chad, Ethiopia and Tanzania. In these countries, mean weight starts to falter at about 3 months of age, and continues to drop sharply until 10 months, with further rapid declines until about 12 months. In all the countries except Chad, there is a slower decline approaching a levelling out at about 35 months. Sharp growth faltering in the early years of life demonstrates the importance of concentrating preventive interventions during this crucial stage of life.

Figure 24 – National samples of growth faltering (weight-for-age)

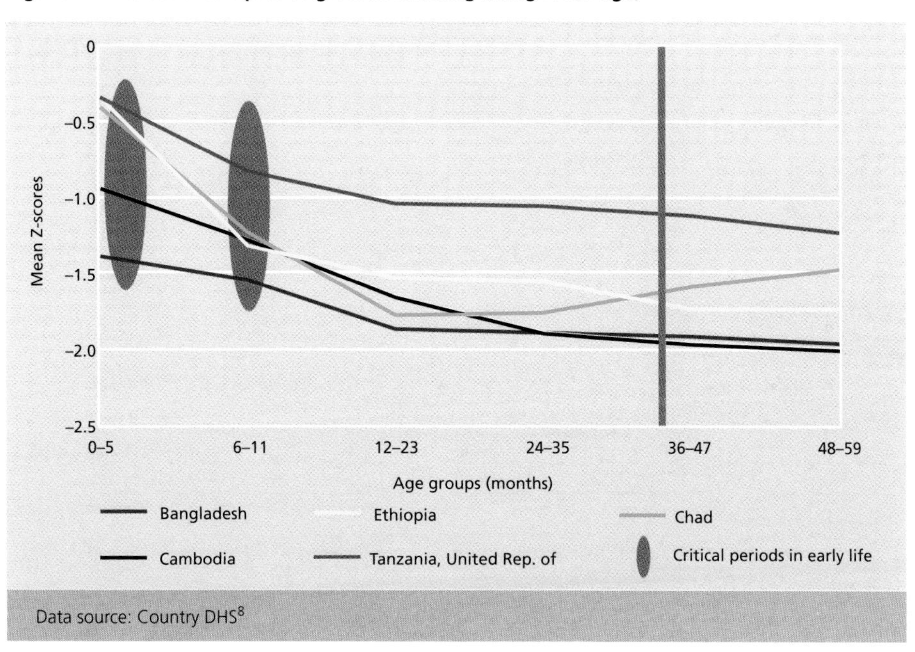

Data source: Country DHS[8]

During pregnancy, especially a first pregnancy, the primary objectives are to prevent maternal mortality and low birthweight babies. Monitoring pregnancy weight gain and child growth is essential for preventing or treating nutrition deficiencies or health problems. Dietary supplementation to ensure sufficient energy intake has been shown to be one of the most important actions, especially during the second and third trimesters of pregnancy. Micronutrient supplementation for pregnant women with anaemia has proven to be particularly effective.

Ensuring access to nutrient-dense complementary foods. Complementary foods are not weaning food, which implies the cessation of breastfeeding. However, semi-solids should be progressively introduced to a young child's diet to complement breastfeeding.

The quantity, quality, form and frequency of use of complementary foods are important. They should be safe, palatable, energy-dense, micronutrient-rich, and prepared from four basic ingredients: (i) cereals or tubers, (ii) protein sources, (iii) vitamin and mineral supplements, and (iv) energy sources such as oil (WFP, 2000).

The introduction of complementary foods is more effective when accompanied by adequate food and childcare practices. A proven "package" includes feeding (breastfeeding and complementary feeding), hygiene and sanitation (related to food preparation and access to clean water), and childcare (stimulation).

Breast milk is more than just food

Breastfeeding protects babies from diarrhoea and ARIs, stimulates their immune systems and improves response to vaccinations. Breast milk, recommended during at least the first two years and exclusively for the first six months of the baby's life, is often replaced by commercial or other substitutes. Formula feeding is expensive and carries risks of additional illness and death, particularly where the levels of infectious disease are high and where preparation and storage of substitutes are not carried out properly. Children living in disease-ridden and unhygienic conditions, who are not breastfed, are between 6 and 25 times more likely to die of diarrhoea, and 4 times more likely to die of pneumonia than breastfed infants (UNICEF, 2007a).

Exclusive breastfeeding for the first six months. WHO and UNICEF recommend exclusive breastfeeding for six months to reduce significantly child mortality and malnutrition, as well as continued breastfeeding with safe, appropriate and adequate complementary feeding until at least 2 years to reduce mortality and stunting. Success depends on promotion and education activities to help lactating mothers improve the effectiveness of breastfeeding practices.

Figure 25 – Practical solutions for pregnant and lactating women, infants and young children

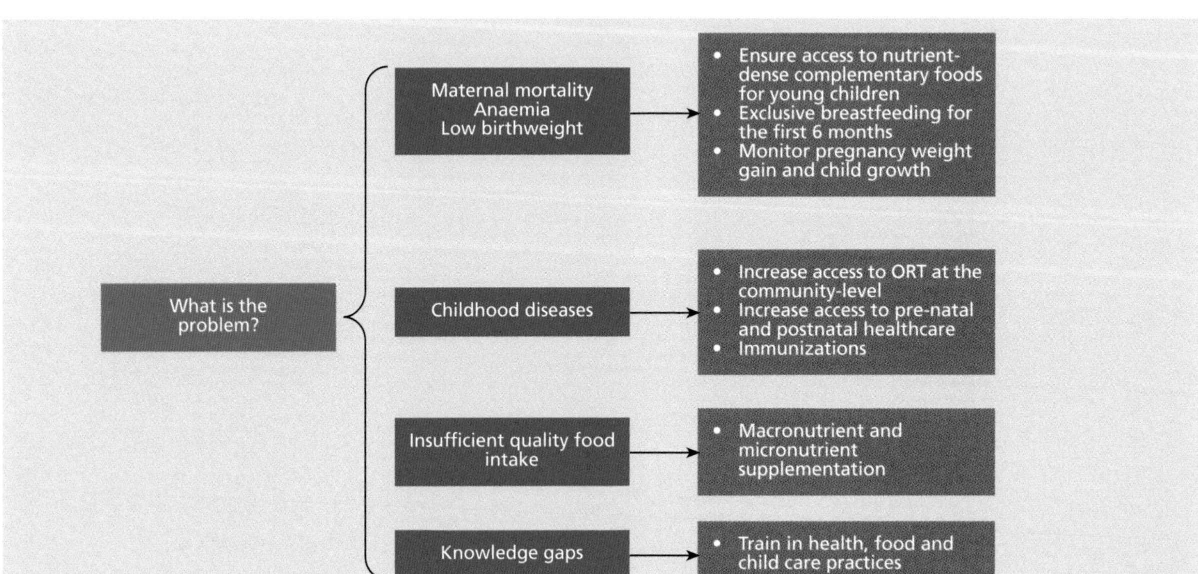

Monitoring weight gain and child growth. A child's growth is very sensitive to nutrition and disease. Problems, particularly undernutrition, can be detected by means of growth monitoring long before the appearance of other signs and symptoms. The most effective way of monitoring growth is to measure weight and height as an integrated component, as is done in mother-and-child health programmes.

Community-based oral rehydration therapy. ORT treats the deficits in water and electrolytes that occur with diarrhoea. Packets of oral rehydration salts (ORS) costing about US$0.10 each or a simple solution of sugar, salt and water would substantially decrease child deaths from diarrhoeal dehydration. Given the effectiveness of ORT and the prevalence of diarrhoeal diseases, community clinics need to be equipped to provide ORT for children at risk (WHO and UNICEF, 2006). For example, ORT can be made readily available at the community level with child immunization.

Immunization. Immunization is one of the most important and cost-effective interventions that health systems can provide; it is essential for saving children's lives. It is an affordable means of protecting whole communities and reducing poverty. Immunization is estimated to have saved 20 million lives in the last two decades (UNICEF, 2007b).

ORT in Bangladesh

Between 1980 and 1990, a Bangladeshi NGO taught over 12 million mothers how to prepare ORT at home with salt and brown sugar. The training was supported by the promotion and distribution of pre-packaged ORT by the Government and various agencies. In 1993, a national assessment found that 70 percent of mothers knew how to prepare safe and effective ORT at home and that ORT was used to treat 60 percent of all diarrhoeal episodes. Drug sellers and village doctors now recommend ORT more frequently than before, and the availability of pre-packaged ORS in rural pharmacies has improved. There is convincing evidence that wide-scale promotion of ORS for dehydration related to diarrhoea has led to these improvements (Chowdhury et al., 1997).

Adolescence: solutions reaching school-age children and adolescents

For school-age children and adolescents, solutions need to address growth and cognitive development, promote a healthy diet and lifestyle, and communicate knowledge to prevent health risks.

Promoting a healthy lifestyle. The most effective interventions – in addition to an adequate quality diet – are those that address social and behavioural challenges faced by adolescents, such as smoking,

Figure 26 – Practical solutions for school-age children and adolescents

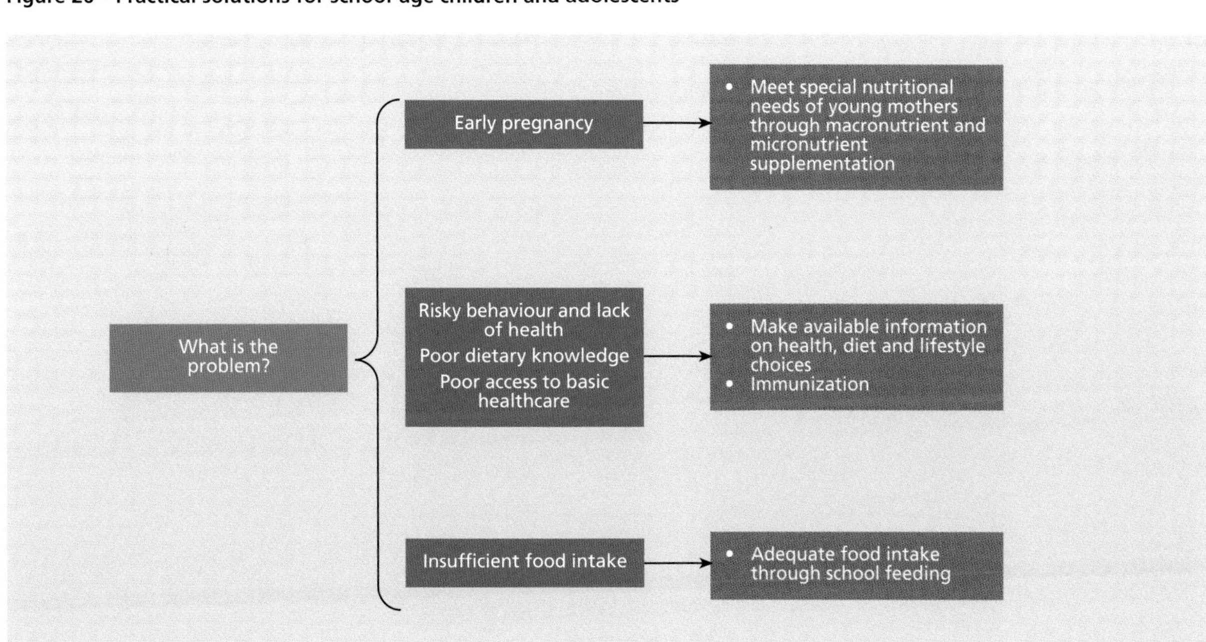

risky sexual behaviour, HIV prevention, and healthy lifestyle promotion. Long-term strategies that improve nutritional status, delay marriage and first pregnancy, keep girls in school and empower young women are also fundamental.

School feeding. In addition to being an incentive for parents to send children to school, school feeding improves nutrition and health, and leads to better performance, with fewer repeated grades and reduced drop-out rates. School feeding accompanied by complementary education activities can address issues such as nutrition and dietary diversity, hygiene, healthy lifestyle and HIV/AIDS prevention.

The benefit of these essential solutions is that they eliminate hunger, result in better health, promote social equality, and increase the effectiveness of other investments in human capital.

Innovative school feeding for adolescents

Older students, especially those of high-school age, generally have lower participation rates in school meals than children in primary grades: breakfast is often perceived as "uncool". Communities of the Flathead Native American Reservation in Ronan, Montana, found an effective way of increasing participation among older students; they offered fresh fruit as part of the programme. Adding fresh fruit was a simple yet effective solution because it allowed students to take an apple, pear or banana and go on their way. As a result, the number of middle-school and high-school students eating fruit in the morning jumped from 10 percent to 70 percent (USDA, 2005).

Intermezzo 7: Sprinkles – an innovative, cost-effective approach to providing micronutrients to children

Micronutrient deficiencies affect a third of the world's population and account for an estimated 7.3 percent of the global burden of disease. Micronutrients comprise fat- and water-soluble vitamins and minerals. Deficiencies of iron, vitamin A and iodine are estimated to be the most prevalent. Among the population groups, the most vulnerable are infants, young children and pregnant women because of their greater physiological needs. Iron deficiency anaemia affects 750 million children worldwide, making it the most frequent micronutrient deficiency; it occurs across high-, middle- and low-income countries. Over 100 million young children suffer from vitamin A deficiency, which leads to partial or total blindness in its most severe form. Iodine deficiency disorders affect an estimated 43 million people worldwide; they are the single most important cause of preventable mental retardation.

In the past, during crisis situations, most emphasis has been placed on relieving hunger and macronutrient deficiencies. More recently, it has been determined that micronutrient deficiencies are just as important, especially in populations where deficiencies existed before a crisis. Providing vitamins and minerals for populations at risk, especially in emergencies, plays a major role in decreasing death, morbidity and susceptibility to infections. Hence, recommendations have recently been made to ensure adequate intake of micronutrients during emergencies (WHO, 2006). "Home fortification" of foods with Sprinkles is an important advance in the global challenge of reducing micronutrient deficiencies in crisis situations and in development programmes.

What are Sprinkles?

In developed countries, commercial foods are generally fortified with vitamins and minerals. Sprinkles are a unique product that provides micronutrients to fortify foods prepared in the home – hence the term "home fortification".

Sprinkles are tasteless, multiple-micronutrient whitish powder, packaged in single-serving sachets. The entire content of one sachet per day is "sprinkled" on or mixed into any semi-solid or homemade food. For example, Sprinkles can be added to fortified or non-fortified food assistance rations such as corn–soya blend (CSB) or wheat–soy blend (WSB). They can also be added to porridge, mashes, gruels made from any grain or starchy roots and tubers, or yogurt. The iron in Sprinkles is encapsulated to prevent it from interacting with food, thereby limiting changes to the taste, colour or texture of the food. The formulation provides the flexibility to mix iron, vitamins A, C and D, B-vitamin complex, folic acid, zinc and other micronutrients.

When to use Sprinkles?

The use of Sprinkles is particularly opportune during emergencies where the micronutrient needs of young children are increased by lack of fresh and diverse foods and high rates of disease and infections. Although blended foods distributed during emergencies are regularly fortified with a mix of micronutrients, they may not be adequate to fully address the needs of infants and young children. Sprinkles can therefore be integrated into existing relief aid programmes to contribute to improving the nutritional quality of food assistance rations offered.

What is the evidence that Sprinkles are effective?

To assess the health impact of Sprinkles, community-based studies were conducted across Asia, Africa, and the Americas involving anaemic and non-anaemic infants and young children. To date, Sprinkles have been successfully used in emergencies in, for example, Indonesia and Haiti, development programmes in Bolivia, Guyana and Mongolia, and school feeding programmes in China. Overall, the evidence has demonstrated that Sprinkles interventions are successful in treating and preventing anaemia, are safe, well tolerated by children and acceptable to caregivers and communities.

In Indonesia, Sprinkles were distributed after the tsunami to 200,000 children aged 6 months to 12 years, demonstrating the feasibility of the recent statement of WHO/UNICEF/WFP on providing micronutrients to vulnerable populations as part of emergency relief (De Pee et al., 2006). In Mongolia, Sprinkles were distributed to more than 15,000 children aged 6 months to 3 years as a component of an integrated nutrition programme that reduced anaemia prevalence from 55 percent to 33 percent. In Cambodia, Ghana and India clinical trials showed that between 50 percent and 65 percent of anaemic infants and young children given

Sprinkles recovered from anaemia (Zlotkin et al., 2001; Agostoni et al., 2006; Hirve et al., 2007).

Data collected globally on acceptability and compliance suggest that Sprinkles were well accepted and used in the communities that received the intervention. In China, Haiti and Pakistan, the perceived benefits of Sprinkles for children reported by mothers included increased appetite and general improvement in children's health, learning capacity and well-being. In Bangladesh, major reasons cited for liking Sprinkles included ease of mixing Sprinkles with complementary foods and that their use promoted the appropriate introduction of complementary foods.

What are the advantages and side benefits of using Sprinkles?
In crisis situations, Sprinkles can be added to any semi-solid food; there is no need to wait for the arrival of fortified food assistance. Recent evidence has demonstrated that giving Sprinkles to young children in addition to fortified WSB was more beneficial in decreasing anaemia prevalence than giving WSB alone (Menon et al., 2007).

Sprinkles are designed to contain the amount of minerals and vitamins needed by children. They can be added to locally prepared foods already being consumed without the risk of children rejecting novel and unfamiliar commercial products. The challenging task of changing traditional feeding habits is therefore minimized. Poor storage and overcooking may sometimes reduce the micronutrient content of fortified food, but Sprinkles are added to foods after cooking, thereby minimizing any micronutrient loss.

Apart from providing micronutrients, Sprinkles can contribute to healthy weaning because they can only be used with complementary foods. The use of Sprinkles does not conflict with breastfeeding and can help to promote timely transition from exclusive breastfeeding to complementary foods at 6 months, as recommended by WHO. The single-dose sachets are convenient and easy to use and do not require special measuring or literacy skills. Minimum storage needs are required, because the sachets take only a small amount of space – each sachet measures 3.5 x 6 cm and weighs 1 g – and have a two-year shelf life. Cost depends on the quantity of sachets ordered, the composition of the mixture and the site of production, but it is generally 1.5–3.5 US cents per sachet. In addition to their beneficial effects and high impact on health and nutritional status, the sachets are lightweight and so easily transported and distributed; they are hence a cost-effective and operationally feasible approach to delivering micronutrients to vulnerable children.

What is needed for implementing a Sprinkles programme?
The success of large-scale distribution of Sprinkles will depend on a well-organized distribution system with a regular supply of Sprinkles, adequate population coverage in relation to the target group and education and motivation of recipients to ensure demand and compliance. A multi-sectoral coalition of partners is necessary which involves dialogue with the ministries of health, United Nations agencies, NGOs, civil-society organizations and the private sector.

The characteristics of a Sprinkles programme will vary between countries depending on available infrastructure. Distribution partners may include public-sector, civil-society and private organizations. In the initial stage of developing a control strategy for Sprinkles, information on the micronutrient status of the target population as well as current and past experiences with micronutrient programmes is invaluable. For cost-effectiveness, it is beneficial to integrate the Sprinkles intervention into ongoing programmes such as health-related or community-based programmes that have the capacity to develop and manage the distribution, evaluate the intervention and link it with their goals.

A challenge in programme implementation involves communication with beneficiaries to enhance adherence and sustainable use. This necessitates culturally appropriate communication and media campaigns to educate beneficiaries on the proper use and importance of Sprinkles and the positive effects on health, especially when Sprinkles are being used for the first time. Promotion of community participation in the Sprinkles intervention should be highly encouraged.

Contributed to the World Hunger Series *by Dr Stanley Zlotkin, Professor, Nutritional Sciences and Public Health Sciences, University of Toronto.*

Map 7 – Health inequalities across the world

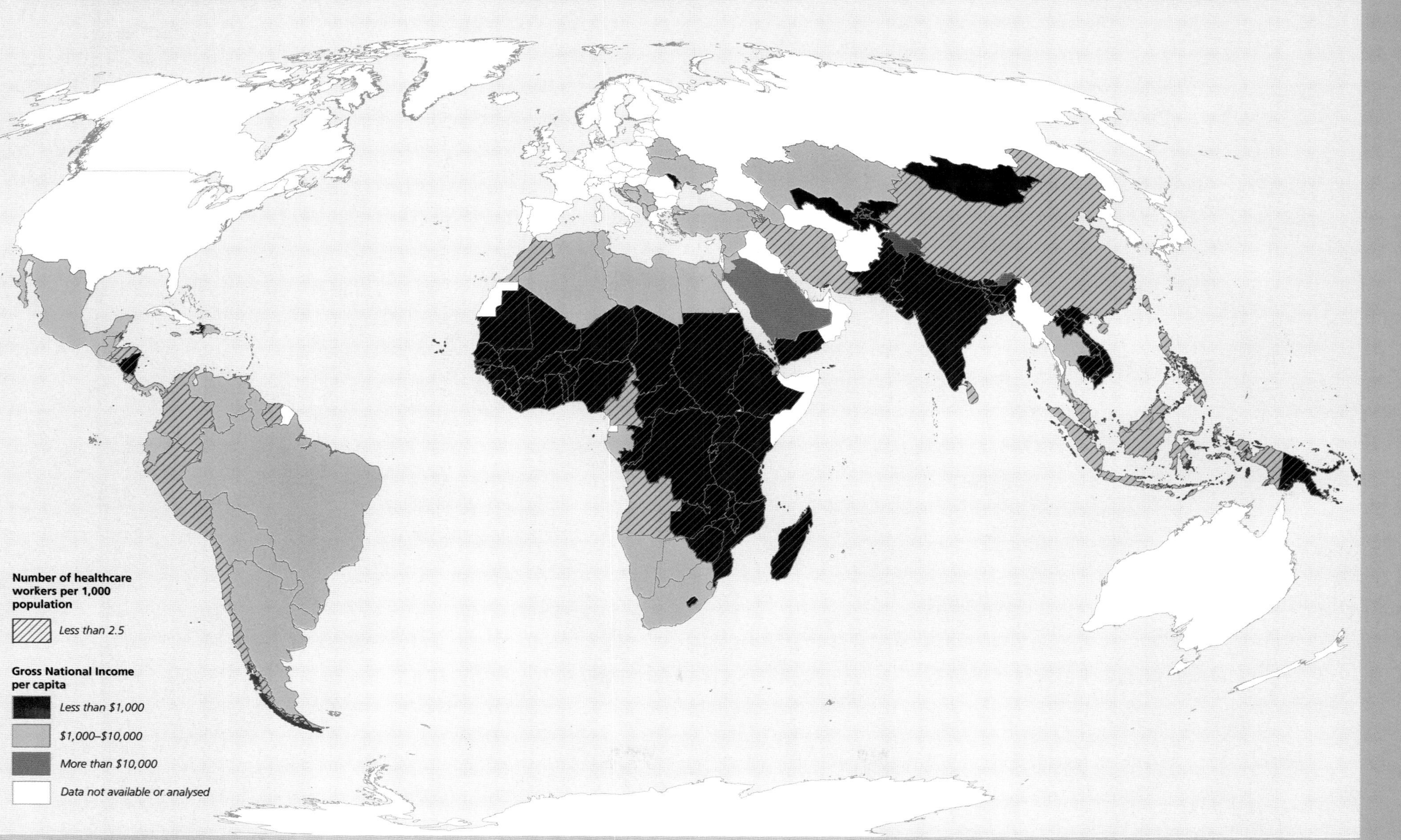

The boundaries and the designations used on this map do not imply any official endorsement or acceptance by the United Nations. Map produced by WFP VAM.

Data sources: OECD, 2006; WHO, 2007

3.3 Making the right political choices

"The way humans grow is a 'mirror' reflecting the socio-economic conditions of the society."

John Komlos and Benjamin E. Lauderdale, 2007[9]

Despite the various cost-effective solutions for combating hunger and improving health, and the potential to direct national and international political commitments to address these related problems for the poorest people, efforts are still insufficient. There is a real risk that the MDGs, themselves relatively modest, may not be met. The *World Hunger Series 2007* challenges leaders to build on past successes, combining current knowledge with a will to undertake practical and effective solutions to end hunger in the coming decades.

There are four strong motivations for prioritizing these hunger health solutions:

- The costs of hunger and poor health are high.
- Solutions are affordable, cost-effective and sustainable.
- There is consensus on the human right to adequate food, nutrition and health for all.
- Well-fed and healthy populations contribute to economic growth more effectively.

The challenge for leaders is to garner commitment and formulate strategies to address hunger and poor health together as explicit targets of human and economic development. This chapter sets out a number of areas where commitment could be strengthened.

Increasing awareness: the heavy burden of hunger

An obvious foundation for effective action is to draw the attention of decision-makers to the empirical evidence demonstrating the burden of hunger, undernutrition and disease for different socio-economic groups. This requires governments, individuals and agencies with expertise in health and nutrition to compile data at sub-national levels and for specific vulnerable groups. It also requires commitment to assess interventions and translate the findings into clear and feasible proposals for policymakers and the public.

Advocacy for the hungry has suffered because there is a lack of public awareness of the true costs of hunger and related poor health. The manifestations of hunger and poor nutrition, though enormous, are largely insidious and invisible. Even when awareness exists, effective action can be constrained by other priorities. The lack of effective action reflects the powerlessness of those who suffer most from hunger and poor health:

- Neither individuals, families, nor governments recognize the full human and economic costs of hunger.
- Governments do not recognize that hunger reduction actions are among the most effective interventions for combating slow economic growth and poverty.
- There is not always consensus on how best to intervene against hunger and poor health.

In the end, commitment determines whether interventions are effective and sustainable. Unfortunately, many past interventions have suffered by being poorly conceived, underfunded and unable to demonstrate impact.

Increasing commitment: eradicating hunger as a priority goal

"If you do not know where you are going any road will take you there."

Lewis Carroll (1865)

The elimination of hunger cannot be relegated as a subsidiary goal of other commitments. In view of the tremendous human, economic and social costs of hunger, its elimination must be a development priority and an integral part of health goals. Aligning hunger and health with other sectoral strategies is critical and they must be included as priorities in national and local development plans. Setting hunger elimination as a goal involves accurate identification of prevailing hunger and health problems, their causes, the social groups most affected and vulnerable, alleviation targets, the assignment of responsibility, and accountability mechanisms.

A number of countries have developed and implemented comprehensive national health and nutrition plans that include explicit priorities to eliminate hunger. These plans recognize that the alleviation of hunger and related morbidity and mortality cannot be achieved by a single intervention. Given the futility of providing general blueprints for progressing towards development goals, it is the role of policymakers to make country-specific decisions on strategies to combat hunger and poor health.

Binding commitments to hunger and health goals in constitutions, national development plans or international agreements are good starting points, provided they include:

- policies that can be translated into targets, targets into action plans, and plans supported by yearly budget allocations that are regularly monitored;
- regular funding for inclusion as an item in national budgets;
- an organizational set-up that provides leadership and administrative capacity for action; and
- assignment of responsibilities to ensure that actions are fully implemented.

However, governments often face competing demands and priorities can change, affecting commitment to hunger and health goals. Donor priorities and trends also influence country priorities; thus donors need to support government plans to reduce hunger.

Maximizing resources

Increasing human and financial resources is fundamental to eliminating the burden of hunger and poor health.

Resource commitments must follow goals and priorities. Strategies to eliminate hunger and promote health will need to overcome resource constraints. To achieve optimal impact, appropriate resources must be allocated and their use maximized. The resources needed are not only financial: they include leadership, management and system support to make social services effective.

Programmes must be designed to be sustainable and a critical element is breaking down inequalities by vulnerable group and gender. A number of regions, countries and sub-national areas have made great strides in reducing hunger and improving health: Sri Lanka and the Indian state of Kerala provide inspiring examples. Conversely, several countries with impressive and sustained economic development continue to be burdened with high rates of food insecurity and undernutrition. This is because income growth alone does not lead to sufficient improvements in nutrition and health, and economic progress does not necessarily "trickle down" to the poorest people.

Increasing government spending: a good start

On a regional basis, expenditures on health as a percentage of government spending increased from 2000 to 2004 in Asia, the Middle East and sub-Saharan Africa. The increases, though slow, correlate with positive trends in the reduction of deaths, especially among very young children (WHO, 2007).

There are many successful country-level programmes in health and nutrition in low- and middle-income

Figure 27 – Government spending on health as a percentage of GDP

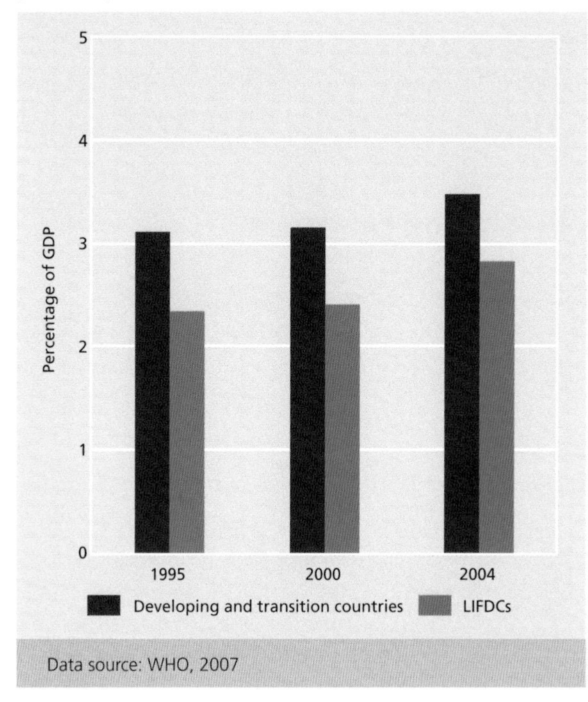

Data source: WHO, 2007

countries: Brazil, Honduras, Sri Lanka and Uganda are examples where national programmes have been administered effectively. Chile's long-term success with national programmes such as the complementary feeding programme are well known: there was a dramatic fall in infant mortality rates over three decades from 119.5 to 16 per 1,000 live births, and life expectancy increased by 13 years (Binswanger and Landell-Mills, 1995).

However, increasing the levels of health and nutrition expenditures is just one aspect, albeit important. Optimizing the effectiveness and efficiency of resource use requires even stronger commitment. There are many areas where governments can maximize resources: financial and administrative management are good examples.

Much can be done to streamline bureaucratic processes to save money and time.

The *Progresa* programme in Mexico offers numerous insights into streamlining and increasing efficiency. The Government put systems in place to promote accountability and transparency, which are critical to maximizing the effectiveness of resources. Social services were expanded and focused to prioritize improved nutrition for very young children, better targeting to reach the most vulnerable and improved institutional monitoring to show results at the impact level (Skoufias and Parker, 2001).

The challenge is for all those involved in nutrition and health to optimize resources to reach hungry people.

Using aid resources more effectively

Although the overall flow of aid resources has increased, the level of funding for basic public health has remained fairly low. Donors have shown more willingness to respond to conflict, natural disasters and single diseases than to ensure aid for basic public health and nutrition, ignoring the fact that these basic investments could reduce the impact of shocks and disasters.

Less than 5 percent of global health research in the 1990s was devoted to diseases and health problems endemic to developing countries; less than 10 percent of health research was directed towards the major problems affecting 90 percent of the world's population. As a result, very few new products for diseases that are mainly endemic in poor countries have been registered for clinical use (Fernholz et al., 2007).

High-profile campaigns for specific diseases have attracted significant additional funding. An analysis of donor funding shows that funds have been skewed heavily towards particular diseases, well beyond the level of their prevalence and impact. It calculated donor funding for 20 historically high-burden infectious diseases for the years 1996–2003 using data from 42 donors, classifying grants according to disease.

The data showed that funding did not correspond closely with burden (Shiffman, 2006). For example, ARIs comprised 25 percent of the burden of disease but received less than 3 percent of direct aid. Malaria also stood out as a neglected disease, with 9 percent of funding but 14 percent of the burden. TB was relatively underfunded, although closer to parity. Other diseases that received high levels of funding with contrastingly low burdens included polio, which received 14 percent of funding but accounted for less

Figure 28 – Government expenditure on health

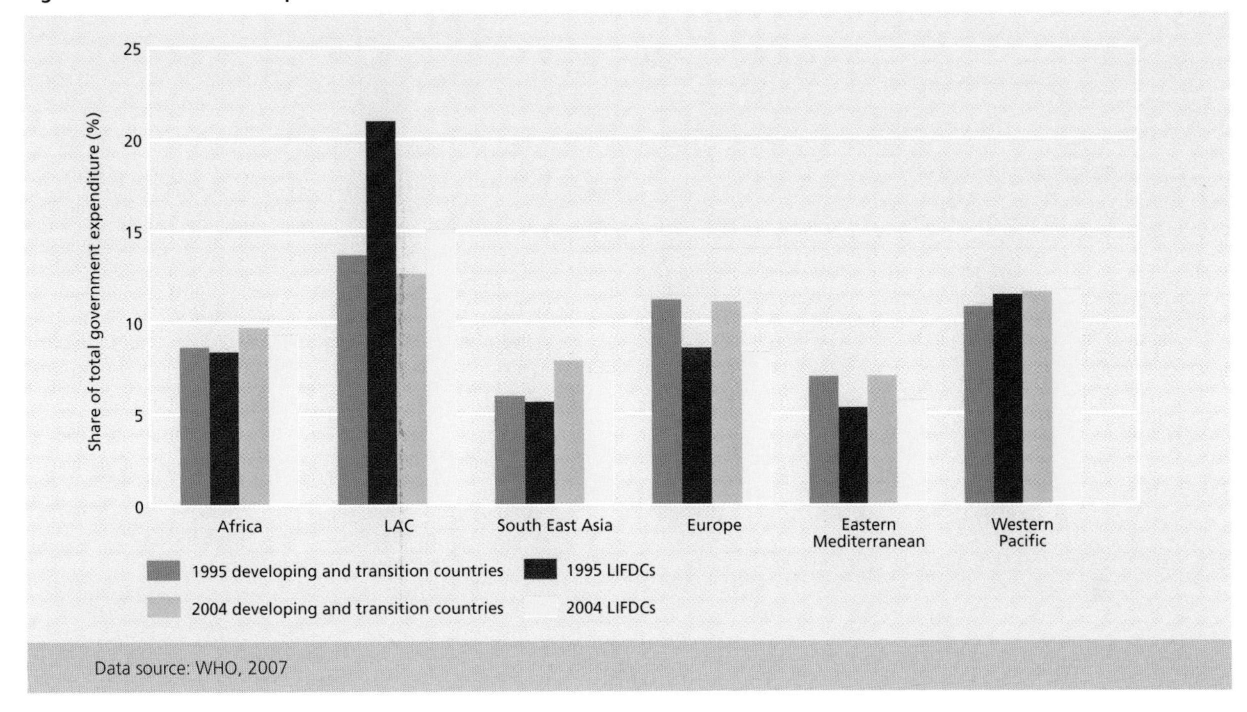

Data source: WHO, 2007

than 0.1 percent of the disease burden. HIV/AIDS received 46 percent of funding and accounted for 31 percent of the burden of disease.

Some valid factors account for this variance, including the high costs associated with the final stage of globally eliminating specific diseases such as smallpox and polio, and a focus on diseases for which cost-effective solutions exist. Resources should be directed to where there is the most need.

New sources of funding

Funding has been influenced by the emergence of pressure groups focused on individual diseases. These groups have helped to increase the overall level of resources, but the diseases that have attracted most donor attention are sometimes of less relevance to developing countries (Fernholz et al., 2007).

Treatment has tended to attract greater funding than prevention – an example is HIV/AIDS – in part because of the possibility of demonstrating positive results. Attention has been paid to perceived "quick returns" rather than to areas that require longer-term interventions. Institutions that have a positive effect on poor people, such as basic healthcare delivery systems and public health improvements – especially water and sanitation – too often remain weak and underfunded.

The recent large influx of funds through philanthropic organizations for addressing health problems such as HIV/AIDS, TB and malaria have given a tremendous boost to the treatment and prevention of these diseases. But they have also raised questions about the effectiveness of interventions in the health sector, which are implemented "vertically" in that they target specific diseases and conditions; "horizontal" approaches focus on improvements in public health services and address the basic causes of hunger and poor health.

Many of the philanthropic funding organizations created in the last five to ten years have concentrated on vertical approaches with a rigorously interventionist approach. The proponents of the vertical approach argue that serious global health problems, especially HIV/AIDS, require emergency-style interventions with their own infrastructures for delivery that focus on a narrow set of measurable outcomes. This may be a valid approach, but the risks include undermining existing health structures by attracting health professionals away from national programmes and diverting funds away from basic nutrition and healthcare.

Migration of skilled healthcare workers

WHO estimates that more then 4 million additional doctors, nurses, midwives, managers and public-health workers are urgently needed in developing countries. The situation is made worse by the migration of skilled healthcare workers. An unbalanced global employment situation means that healthcare professionals, especially nurses, are being drawn away from poor countries to rich countries. The Commonwealth Code of Practice for the International Recruitment of Health Workers is currently the only international code to try to rectify this situation. There has been a vast expansion of education, training and capacity-building worldwide, but the very people needed to support health programmes are decreasing in number: village health workers, nurses and nutritionists in ministries of health and agriculture. The people who desperately need education – women and adolescent girls – are often overlooked (WHO, 2006).

Pattern of funding partnerships

A recent review of the emerging pattern of funding concluded that new funding partnerships bring:

- international attention to health issues, putting them on national and international agendas;
- additional funds for basic health programmes and new research;
- improved international funding standards and norms;
- improved access to cost-effective healthcare interventions for people with limited ability to pay;
- strengthened national health policy processes and content with a focus on outcomes; and
- augmented capacity to deliver basic health services.

Source: Buse and Harmer, 2007

Much more could be achieved if better use were made of available resources. Increased domestic and external financing will be required if the MDGs are to be met – and even more if the MDGs are to be surpassed.

Making partnerships work

Long-term service delivery requires strengthening partnerships. In March 2005, the Paris High-Level Forum followed up the Rome Commitments with a declaration on aid effectiveness covering ownership, harmonization, alignment, results and mutual accountability. Harmonization is ongoing, but the transaction costs can be high because of the difficulties of implementing programmes under a single budget or with shared personnel.

Coordination and harmonization require the involvement of numerous stakeholders in food security, nutrition and health at the national and community levels. Support for participatory and country-led development planning based on greater sharing of information and support for community-based initiatives is a critical component of nutrition and health service delivery.

Over the last ten years, national and international NGOs have increased their commitment to fighting hunger and poor health. The capacity of NGOs to advocate for resources for causes and to work effectively among local people is widely recognized. They have been particularly effective in campaigning for and securing resource commitments from government and private sources for development.

Partnerships are fundamental to eradicate hunger and poor health. No single government or organization can hope to do so alone. Partnerships should include national and local governments and the people being served. They should be based on the principles of cooperation and complementarity: partnerships are effective when there is a shared understanding of the desired outcomes and how to achieve them, and when complementary inputs and skills are contributed.

It takes a dedicated effort to establish partnerships that will make a difference.

Commitment to the hungry

In poorer countries, the essential issue is to ensure that government interventions reach vulnerable people in a sustainable manner. One reason why much international development assistance fails to reach the most vulnerable is that they lack a voice and have no power to rectify deficiencies or draw attention to the issues that are most important to them (Easterly, 2006). Nevertheless, there have been recent encouraging signs of change.

The right to food is referred to in the constitutions of some 20 countries. These references are usually unambiguous. For example:

> *"It is the duty of the family, of society, and of the State to ensure children and adolescents, with absolute priority, the right to life, health, food, education, leisure, professional training, culture, dignity, respect, freedom, and family and community life, in addition to safeguarding them against all forms of negligence, discrimination, exploitation, violence, cruelty and oppression"* (Constitution of Brazil, Article 227).

> *"The State shall regard the raising of the level of nutrition and the standard of living of its people and the improvement of public health as among its primary duties ..."* (Constitution of India, Article 47).

These statements are evidence of increasing commitment by governments to protect their citizens from hunger. Yet there are few legal mechanisms to enforce these commitments. Too often, the hungry lack the power and resources to challenge the state to adhere to its commitments.

A handful of countries have made serious efforts to meet their commitments through national hunger eradication programmes that aim to give the poor a more powerful voice. For example Brazil launched *Fome Zero* (Zero Hunger) in 2003, a federal strategy to ensure that poor people have access to food. By June 2006, the programme had reached 11.1 million families.

The flagship component is the *Bolsa Familia* (Family Stipend) programme, through which small monthly cash payments are made to mothers on a sliding scale. This increases the ability of households to purchase the food they most need and value. Targeting mothers also increases the utility of cash transfers: a number of studies show that mothers allocate scarce resources to the best of their ability for their children's well-being (WFP, 2000).

Coordination at the local level is important to ensure the participation of marginal and disenfranchised groups. Various approaches to enable all stakeholders to participate have been developed, though they are not always used. It is important to consider power imbalances and lack of access to information and to understand why some groups are marginalized, with poor access to food and basic health services.

Participatory processes enable different groups to voice their concerns or express support. Decisions can range from prioritizing interventions that form basic health and nutrition services to selecting criteria for identifying recipients. Informal structures exist in all societies; they can serve, for example, to provide rapid feedback and information for monitoring.

The most effective changes have been achieved where it has been demonstrated that improving the welfare of marginalized groups improves the welfare of a whole society. In the case of indigenous peoples, several models and programmes now exist to address poor nutrition in culturally appropriate and socially inclusive ways.

There needs to be ownership and buy-in at all levels – international, national, community and individual. Programmes and projects may take longer to start up, but they have a greater chance of success.

Intermezzo 8: Partnerships to overcome child undernutrition in Latin America and the Caribbean

Constitutional entitlements to health and nutrition in the region

Increased awareness about hunger has motivated Latin American and Caribbean governments to strengthen existing constitutional entitlements to health, food security and nutrition. The constitutional entitlements are increasingly accompanied by more effective national health and nutrition plans – often connected to regional approaches with international support – and the inclusion of hunger in national political agendas.

The right to health is constitutionally stipulated in 17 of the region's 33 countries. The right to food – without which the right to health cannot be implemented – is specifically guaranteed in the fundamental laws of Ecuador and Nicaragua and in the constitutions of the Dominican Republic, Guatemala, Panama and Suriname. In other countries, governments have a constitutional obligation to protect the nutrition of vulnerable groups such as pregnant and lactating women, children, indigenous populations and the very poor. Constitutional provisions cannot by themselves be expected to solve the region's hunger problem, but they provide legal claims on national and international resources and political mandates and procedural channels to help to eliminate hunger in the region.

Regional initiative: towards the eradication of chronic childhood undernutrition in Central America and the Dominican Republic

Current trends towards hunger eradication include regional approaches based on national and international partnerships. One such initiative is "Towards the Eradication of Chronic Childhood Undernutrition by 2015" in Central America and the Dominican Republic (ECLAC and WFP, 2007). This plan, jointly proposed by WFP and the Inter-American Development Bank (IDB), was presented at a June 2006 Regional and Technical Consultation held in Panama. It is particularly relevant because environmental, social, cultural and economic factors continue to hinder Central America's progress towards the MDG hunger goal. In this sub-region, 1 million children from birth to 36 months of age exhibit serious height and weight deficiencies indicative of chronic undernutrition.

"Towards the Eradication of Chronic Childhood Undernutrition by 2015" brings together stakeholders from Belize, Costa Rica, the Dominican Republic, El Salvador, Guatemala, Honduras, Nicaragua and Panama, WFP, IDB, UNICEF, PAHO, the Nutrition Institute of Central America and Panama, and Mexico's National Public Health Institute. The initiative seeks to accelerate the reduction of chronic undernutrition in each of the participating countries, in particular by

Constitutional Provisions on Food Security and Nutrition in Latin America and the Caribbean

"Nicaraguans have the right to be protected against hunger. The State will promote programmes that assure an adequate availability and fair distribution of food."

Constitution of Nicaragua, Article 63

"The State will see that the population's nourishment and nutrition meet minimum health requirements. The specialized institutions of the State will coordinate their actions amongst themselves or with international health organizations, to achieve an effective national food system."

Constitution of Guatemala, Article 99

"With regard to health, the following activities correspond primarily to the State, which shall integrate in them preventive, curative, and rehabilitative functions: 1) Developing a national food and nutrition policy that ensures an optimal nutritional state for all the population, by promoting the availability, consumption and biological utilization of adequate foods."

Constitution of Panama, Article 110

concentrating on pregnant women and children under 36 months. It provides motivation to increase public investment in preventive interventions and to improve their effectiveness. Objectives include developing common criteria for national hunger reduction programmes, supporting the formulation of national plans for reducing chronic undernutrition and giving more visibility to decision-making entities and best practices through regional consultations to present results.

Government responses in Central America and the Dominican Republic

Government representatives from the countries involved reiterated their support for the regional initiative and committed to the formulation of national plans to eradicate chronic undernutrition. Health ministers confirmed the importance of linking the proposal to national and regional emergency-response mechanisms.

Broad regional support was followed by related specific national policy development in the individual countries of the sub-region. In December 2006, Costa Rica released its 2006–2010 Nutrition and Health Policy, which explicitly calls for the development and implementation of the plan "Towards the Eradication of Chronic Childhood Malnutrition by 2015". Guatemala launched a comparable "Chronic Undernutrition Reduction Programme" in 2006 that includes a hunger cost analysis. In El Salvador, agencies cooperated with the Government to draft a 2007–2012 national nutrition plan to further the achievement of the MDGs.

Honduras' nutrition plan, currently being prepared, will be a framework for a national plan that calls for specific child-nutrition interventions. In Nicaragua, a national programme is being developed to be incorporated into the social agenda. The Dominican Republic launched its national plan to eradicate chronic undernutrition as well as the cost-of-hunger report.

"Towards the Eradication of Chronic Childhood Undernutrition by 2015" and the national efforts promoted by the initiative received further endorsement from the finance ministers of the countries involved. Meeting in Guatemala in March 2007 during the 48th annual reunion of IDB governors, the ministers agreed to place infant nutrition at the highest level in poverty-reduction strategies and to accelerate actions toward the eradication of chronic undernutrition (ECLAC and WFP, 2007).

The way ahead

In Central America and the Dominican Republic, the United Nations system has adopted a united approach to support national government plans to eradicate chronic childhood undernutrition; joint United Nations programmes are under way in eight countries. South–South cooperation, such as that between Chile and Peru and between Chile and El Salvador, has great potential in the fight against hunger. These activities provide an opportunity to share best practices and successful experiences among countries engaged in combating undernutrition.

Contributed to the World Hunger Series *by the Latin America and the Caribbean Regional Bureau (ODP), WFP.*

Commitment to what works

A major challenge is to learn lessons from successful hunger and health initiatives and to replicate them in other places and cultures. To scale up activities, it is important to measure results and to know what works. Subsequently, resources need to be allocated to projects that achieve impact.

The benefit of scaling up or replicating proven programmes is that they lead to increased probability of long-term success. Different stakeholders can bring unique resources to a scaling-up process, and together they often create the conditions that lead to further expansion and sustainability.

Resources need to be reallocated to solutions that achieve results; additional resources must be secured to bring projects to scale.

International development assistance is an important source of funding for social programmes, especially in poorer countries. In the last five years there has been a gradual rise in the volume of resources provided by the international community for global health and hunger reduction programmes. According to the Organisation for Economic Co-operation and Development (OECD) Development Assistance Committee (DAC), total official development assistance (ODA) rose by 7 percent in real terms in 2002 and by a further 5 percent in 2003 to its highest level ever in nominal and real terms. On the basis of donors' commitments, continued growth in ODA is expected (OECD, 2007).

The power of measuring results

If you do not measure results, you cannot tell success from failure.
If you cannot see success, you cannot reward it.
If you cannot reward success, you are probably rewarding failure.
If you cannot see success you cannot learn from it.
If you cannot recognize failure you cannot correct it.
If you cannot demonstrate results, you cannot win public support.

Source: adapted from Osborne and Gaebler, 1992

Exciting innovations are emerging in the area of nutrition and health with the establishment of private foundations that pledge high levels of resources to support initiatives. Two such foundations in recent years are the Bill and Melinda Gates Foundation and the William J. Clinton Foundation, both of them placing high priority on health issues and attracting other leaders to join them. These foundations provide financial resources, critical managerial guidance and technical resources for programmes and initiatives all over the world. They support approaches such as public/private–NGO partnerships, and are a significant source of financing for pioneering initiatives, and for scaling up targeted programmes.

Figure 29 – ODA and food aid (2000–2005)

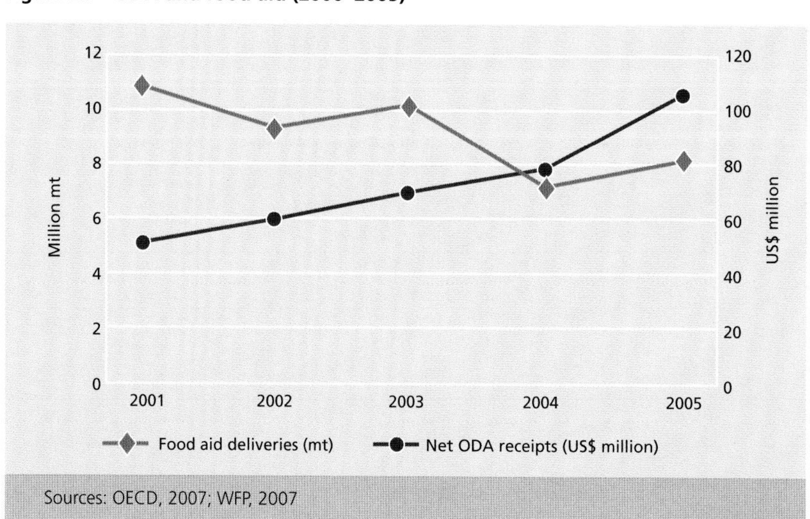

Sources: OECD, 2007; WFP, 2007

The private sector is an important actor that brings in resources for scaling up programmes. The challenge is to recognize the framework in which they operate and to invoke their sense of corporate responsibility. Private companies can provide leadership, skills and financial and material resources.

A range of products can be developed with the help of the private sector. For example, the private sector fills a niche in the area of fortification of salt with iodine and the use of rice and wheat flour to carry vitamin A and iron. Addressing hunger through programmes such as school feeding provides opportunities to involve local companies in the provision of food and services such as transport.

New information technologies open up opportunities in the field of health and nutrition. Health clinics equipped with computers linked to the internet can serve as centres for training and the exchange of ideas. More rapid communication is possible among stakeholders, and monitoring of field activities can be more accurate and timely.

Scaling up successful programmes can achieve a larger impact; but it requires learning from experience. Some general lessons include:

- support local structures;
- develop a commitment to alignment and harmonization;
- strive for balanced representation of stakeholders;
- practise mutual accountability and set realistic targets;
- monitor and evaluate programmes; and
- incorporate learning and innovation into programme implementation.

The choice is ours

These proven and practical essential solutions address the interrelated causes of hunger and poor health. They can increase the effectiveness of other investments in human capital and improve social equity. As demonstrated in Part III, these solutions are also cost-effective. We need to mobilize our collective will to make the right choices. The costs of inaction are high – economically, politically and, most importantly, morally.

Intermezzo 9: From research to action

Malawi has a population of 12 million, of whom 2.5 million are under 5. Of these, an estimated 48 percent are chronically undernourished and 22 percent acutely malnourished. In Malawi, severe acute malnutrition (SAM) is the most common reason for paediatric hospital admission. In the past, mothers had to leave their families for up to a month to travel to the nearest town with a severely malnourished child for treatment. Given the high costs to the family of a prolonged maternal absence, severely malnourished children arrive very late, if at all, at crowded inpatient units. Delayed arrival is associated with more complications and frequently results in inpatient mortality rates above 25 percent.

Community-based therapeutic care (CTC) is a new approach for the treatment of SAM that allows children to be treated at home rather than in hospital. The outpatient model is delivered through rural primary health centres or clinics and was first implemented in Malawi in 2002 through Concern Worldwide programmes in Dowa district. CTC consists of (i) exclusive outpatient treatment for severely malnourished children without complications using energy-dense, micronutrient-rich food, and (ii) a combined approach for the more sick children, whose treatment starts with a short period in a stabilization centre as an inpatient before early discharge into outpatient care. Results have been impressive from the start: recovery rates of 75 percent exceed international standards for therapeutic feeding programmes. Since 2003–2005, CTC coverage has consistently been above 70 percent in Dowa.

Encouraged by these initial findings, researchers started a trial at one of Malawi's main hospitals, where inpatient mortality from SAM at the specialized therapeutic feeding unit had been over 20 percent for decades, despite attempts to bring the rate down. The aim was to start discharging at a much earlier stage so that treatment could be completed in the community, thereby reducing overcrowding and subsequent cross-infection rates and reducing the cost to caregivers. Although cure rates at the therapeutic unit rose from 45 percent to approximately 60 percent, overall mortality remained above international standards. Persistently high mortality resulted from the very high HIV prevalence among children with SAM in urban Malawi and the late arrival of many children for treatment.

New research in Dowa district showed that 59 percent of HIV-positive children recovered their nutritional status using CTC protocols, despite not using ART. This suggests that in these children severe undernutrition was not attributable to HIV infection alone. The impact of such findings is potentially huge: it suggests that there is a group of HIV-positive children who can be kept alive in the short term using nutrition inputs only, thereby buying them time until they can also access an ART programme.

Equally important, the Dowa research revealed that CTC could act as a good entry point into communities for offering voluntary counselling and testing (VCT). The stigma of HIV in many parts of Malawi has made HIV testing programmes difficult to implement. In Dowa, the introduction of VCT into an existing CTC programme that was trusted by the local population greatly improved their trust in VCT. This, combined with confidence that the health service could offer tangible benefits, produced an uptake of testing of 94 percent for children and 60.7 percent for adult caregivers.

Work to identify the appropriate nutritional support strategy for symptomatic HIV/AIDS-infected adults took place in Salima and Nkhotakota districts in 2005, with support from two local home-based care organizations. A daily ration of 500 g over three months increased the weight and strength of many of the participants to a far greater extent than in previous trials that used less energy-dense blended foods. The recipients liked the food: only 4 of the 72 participants were not able to eat it; the majority regained functional improvement, with 78 percent becoming strong enough to walk to medical services at the local facility. This increase in energy and vitality gave renewed hope.

As a result of nutritional rehabilitation in a short time and of the ease with which ready-to-use therapeutic food can be targeted to sick individuals, the cost per recovery was 75–90 percent cheaper than equivalent blended food programmes. This programme shows the potential for CTC and ready-to-use therapeutic food for rehabilitating a large proportion of malnourished HIV-infected people, increasing the uptake of VCT and allowing previously terminal patients to reach local clinics and access anti-retroviral medication that could extend their productive lives for decades.

Contributed to the World Hunger Series *by Valid International.*[10]

Part IV The Way Forward: Towards a World without Hunger

Governments must reinforce commitment to surpass the MDGs, and eradicate hunger, in part by ensuring access to quality healthcare for the hungry and marginalized.

Part IV: The Way Forward presents ten key actions that support the implementation of the essential solutions presented in Part III. These actions help reinforce commitment to hunger eradication and better health for poor people; they also promote more effective use of knowledge, experience, technology and resources. The Way Forward rests on the premise that hunger and poor health must be addressed together.

4.1 The way forward: ten key actions

"Knowledge is of no value unless you put it into practice."

Anton Chekhov (1860–1904)

Relative decline in height in the United States

"Americans were tallest in the world between colonial times and the middle of the 20th century, have now become shorter (and fatter) than western and northern Europeans. In fact, the US population is currently at the bottom end of the height distribution in advanced industrial countries.

Yet US physical stature does not fully reflect US affluence. Although on paper the United States is still among the wealthiest of countries, its population has become shorter than western and northern Europeans physically and also lives shorter lives.

Why US heights declined in relative terms remains a conundrum, a topic for future research, but even at this stage of our knowledge we can conjecture that there are differences in the diets of US and European children that could affect human growth. American children consume more meals prepared outside the home, more fast food rich in fat, high in energy density and low in essential micronutrients than do European children.

Moreover, consideration of the differences in the socio-economic institutions of Europe and the United States might help in at least beginning to resolve this paradox. Without claiming to propose a comprehensive answer to this quandary, we propose the hypothesis that there are several crucial differences between western and northern European welfare states and the more market oriented economy of the United States that might well shed further insights into this paradox. This includes greater socio-economic inequality and more extensive poverty in the United States. Furthermore, the European welfare states provide a more comprehensive social safety net, including universal healthcare coverage, while the share of those who have no health insurance in the United States is about 15 percent of the population. Is it possible that the western European welfare states are able to provide better healthcare to children and youth ..." (Excerpt: Komlos and Lauderdale, 2007).

Moving towards the next transition

This *World Hunger Series 2007* publication provides evidence that hunger and poor health are solvable problems. It also shows that progress is uneven, in developing countries in particular but in advanced wealthy countries as well. The fluctuations in stature over time in the United States and Europe are an indication that economic progress does not necessarily represent equitable gains in health and quality of life for all population groups. The current nutrition transition illustrates that progress can generate new problems and that economic growth, technology and knowledge are not sufficient to bring all people out of hunger and poor health. Political will and commitment must be garnered to maximize the potential benefits of economic growth and current knowledge.

This section proposes a way forward: ten key actions to overcome hunger and improve health with a view to more effective use of resources and greater impact for the most vulnerable and marginalized people. Each action will in itself generate significant movement towards the eradication of hunger and improvement of health. Implementing several of these actions will generate a greater positive impact. The preferred option is to incorporate all ten actions in national development plans in support of the MDG targets. Powerful interests will probably have to be either brought on board or challenged to ensure that the intended programmes reach those who need them most.

Action 1: Tackle hunger and health together with a focus on the largest health burdens and poorest people

The hunger target is critical for achieving the health-related MDGs as hunger and health are intimately related. Health interventions that ignore the issue of hunger will have less impact and will prove more costly in the long term. Conversely, hunger interventions that pay attention to health issues will address many interrelated causes and achieve greater efficiency and impact.

The proposed essential solutions discussed in Part III emphasize:

- addressing common underlying factors;
- combining strategically the resources and tools at hand, including food and non-food resources; and
- scaling up what works.

If programmes are built around the linkages between hunger and health, they will address interrelated problems in a more holistic way.

Tackling hunger and health together does not deny a role for specific, targeted interventions, for example improved treatment for malaria. Such interventions in the past have had a significant impact in reducing disease and saving lives. But targeted interventions should not be imposed as some sort of universal remedy: programmes should be designed and implemented with the objective of creating synergies, with the role of hunger eradication and basic health clearly delineated among broader objectives. Narrow professional interests cannot be allowed to detract from seeking common approaches.

A greater proportion of resources must be allocated to the problems that affect the greatest number of people. Undernutrition related to diarrhoea and ARIs, for example, is a major cause of poor health and death among children in developing countries, hence a large share of resources should directly address these two diseases. Funding should follow the well-established public health principle of saving the most lives.

Action 2: Target assistance to address critical life cycle stages

Optimal levels of macronutrients and micronutrients are essential at all stages of the life cycle. Nevertheless, Part III proposed two windows of opportunity, demonstrating that assistance will have a major impact on long-term health and well-being, first during early life (including pregnancy and in the first 36 months after birth), and later during adolescence (including school-age children). Solutions should be prioritized during these critical life cycle stages.

After exclusive breastfeeding for the first six months, the introduction of energy-dense complementary foods with the appropriate combination of macronutrients and micronutrients is crucial to prevent growth faltering. Experience from mother-and-child health and nutrition (MCHN) programmes has shown that food support is relatively easy and cost-effective, especially when targeted to pregnant and lactating women and young children. But such programmes are often constrained by lack of resources.

School feeding interventions are a simple and cost-effective way to reach school-age children and adolescents. They help to retain children, particularly girls, in school and improve diets. Schools are an important conduit for transferring knowledge related to health and diet. However, limited resources mean that school feeding generally targets only the primary-school age group.

Action 3: Focus on prevention in addition to treatment

Policymakers must establish basic preventive services in a more comprehensive way. Such services may include:

- promotion of food and childcare practices such as breastfeeding, hygienic food storage and preparation, family planning and promotion of disease prevention;
- provision of basic care, including prenatal care, midwife-assisted deliveries, neonatal care, use of oral rehydration salts, immunization, basic drugs,

systematic deworming and control of malaria in endemic regions;

- increasing local food supply and diversity;
- improving water and sanitation in marginalized communities; and
- awareness raising and knowledge transfer.

Prevention measures such as immunization and the use of bed nets to prevent malaria have the advantage of creating a sense of ownership among beneficiaries and building support for local healthcare systems.

There is increasing evidence that nutrition and food support accompanying treatment for TB, HIV and other infectious diseases increases adherence and improves outcomes, particularly for the poor. This support should become an integral part of treatment programmes, with research accelerated to improve the effectiveness of food and nutrition support aligned with treatment.

Action 4: Improve the micronutrient content of local foods

The pervasive problem of micronutrient deficiencies shows that calories alone are not sufficient for good health. There is a need for increased awareness and understanding with regard to the value of micronutrients throughout the life cycle.

Fortification of food occurs in a number of countries, but more needs to be done. Multiple-micronutrient fortification of commonly consumed products and/or supplements may be a cost-effective strategy to address multiple deficiencies among school-age children, adolescents, refugees and IDPs. Also, more consideration should be given to home fortification.

Action 5: Promote dietary diversity

Millions of people do not consume fruits, vegetables or milk products. This is a leading cause of hidden hunger. Improving dietary diversity to combat micronutrient deficiencies at all stages of the life cycle is critical to improving the overall health of an individual.

A multi-sector approach and innovative strategies are needed to help communities to develop diverse diets. These could consider:

- involving school children in school gardens to improve diet and develop food-production skills;
- community ownership and use of natural resources for expanding local food production;
- minimum environmental standards to guarantee biodiversity and conservation of local food sources; and
- mainstreaming and integrating biodiversity conservation and rural development in health and hunger agendas.

Action 6: Support national objectives and priorities

Whenever possible, donor interventions should support national objectives and priorities through existing government structures. Addressing hunger and poor health requires a long-term approach, as illustrated in Thailand, described in Intermezzo 10. Interventions that fund and rely on parallel structures are generally not sustainable and should be created only as a measure of last resort. Supporting national priorities and strengthening local institutions is an investment to increase capacity to address hunger and poor health, leading to greater long-term impacts.

Donor interventions often bypass government structures on the grounds that too little of the aid channelled through them effectively reaches the poor. There are two reasons for this:

- lack of national capacity to manage own resources and aid effectively; and
- corruption, which is a problem in a number of developing countries.

While these are valid concerns in some countries, they are not in all. Quite simply, interventions that fund and rely on parallel structures and bypass national priorities usually do not help to develop sustainable solutions.

Action 7: **Give a voice to the hungry**

The hungry must be given the chance to identify their own needs, and provided with information to make choices that suit their circumstances.

Attention to creating and maintaining social cohesion is critical. Hunger-reduction and health programmes need to pay particular attention to the cultural realities of the vulnerable and marginalized in order to secure equitable health outcomes for them.

Including the right to food and health in national constitutions is a good start, as is greater use of participatory approaches in the design and implementation of health and hunger programmes. Ultimately, the goal must be to empower the hungry and marginalized, including indigenous people, refugees and IDPs, allowing them control over their own lives.

Gender disparities that discriminate against women and restrict their economic contribution impose high social and economic costs on individuals, households and societies and impede their advancement; such disparities also have a negative impact on assistance interventions. There is a need to work towards an environment that is supportive of education and life-skills training for adolescent girls and the empowerment of women. Efforts to support women need to be accompanied by the message that such efforts have a positive effect on households and societies as a whole.

Action 8: **Improve accountability and programme monitoring**

Clear lines of responsibility and authority help to ensure that corrective actions are taken promptly. Officials must be held accountable for their actions – or inaction. Authority must be earned – and it comes with a cost.

Those who depend on government services will be better served if they know which official is responsible for ensuring that a service functions correctly. Similarly, donors will be more likely to support government services if there are clear lines of accountability.

Monitoring must move closer to those who have the most severe need, and be undertaken at the sub-national level in a consistent manner that allows tracking of progress, for example towards the MDGs. Interventions should be accurately and regularly monitored to ensure that:

- they are functioning effectively;
- intended benefits are reaching the targeted beneficiaries;
- intended benefits are appropriate and have the desired effect;
- remedial action is swiftly taken when problems are identified; and
- data are disaggregated by gender and vulnerable group to capture trends at the local level.

National programmes need to include concurrent nutrition surveillance, monitoring and evaluation. These tools need not be rigid; rather they can be built with flexibility using evolving information technologies that can serve programme implementation and policy formulation.

Action 9: **Prevent mass movements of people**

Attention to hunger as part of early warning, relief and recovery efforts can help to prevent large movements of people – the main cause of disease and death in crisis situations.

Knowledge of the initial nutritional condition of vulnerable populations is critical in averting large-scale crises. Nutritional surveillance as part of crisis prevention and disaster mitigation can help to pinpoint geographic areas vulnerable to food and nutrition crises, to target assistance earlier and reach those in need more efficiently.

Action 10: **Promote hunger and health knowledge**

Transferring knowledge can be more complex than transferring a service. A service can be delivered with a defined input and measured with a corresponding output. Knowledge transfer requires individuals to assimilate new information and to change their behaviour accordingly. Governments and project managers cannot control personal hygiene, the size and composition of meals or visits to health centres.

An emphasis on training and a knowledge-based approach is inherently more permanent and sustainable from one generation to the next. It is crucial that knowledge about diet, food preparation, lifestyle and health is readily accessible, in particular to mothers, young children and school-age children, adolescents and marginalized people. Decision-makers and leaders also must have the latest research and information at their disposal.

Knowledge and key messages can be tailored to specific groups, for example women, adolescents and indigenous peoples so that the elimination of hunger becomes a goal for the individual and family as well as for politicians.

Educating girls is one of the most important investments any family, community or government can make. An education can make a tremendous difference to a woman's ability to raise a healthy, nutrient-secure family and prevent the spread of disease.

In conclusion

Current solutions are not equitably reaching many in need, thus urgent action is needed if hunger is to be eradicated in the coming decades. The burden of hunger and poor health and its effect on national development can be only part of the rationale for acting. Action must address the human suffering caused by hunger and poor health and remove the divide between those who have access to sufficient quality food and healthcare, and those who miss these most essential ingredients for human well-being. Commitment to surpassing the MDGs, eradicating hunger and providing access to quality healthcare for hungry and marginalized people is the only option.

This final part of the *World Hunger Series 2007: Hunger and Health* started with Anton Chekhov's ever relevant words: "Knowledge is of no value unless you put it into practice". This edition consolidates current knowledge about hunger and health, and presents learning and experiences from around the world. Significantly, as a way forward the report emphasizes essential solutions that are proven and cost-effective, supported by ten key actions to help leaders, communities, families and individuals more equitably address hunger and poor health.

In a world of wealth, knowledge and expertise, the world community must work to ensure that the next economic and nutrition transition ends hunger. This can be achieved in the coming decades by ensuring access to quality food that results in a quality life for all. As Eleanor Roosevelt said nearly 100 years ago:

"The freedom of humankind, I contend, is the freedom to eat."[11]

Intermezzo 10: Nutrition a priority in Thailand

Political commitments

Thailand's first National Food and Nutrition Plan was launched in 1977. It recognized that under-nutrition was a multi-faceted problem that required a multi-sector approach, and was managed by the National Food and Nutrition Committee, which represented the ministries of agriculture, education, health and the interior/community development. A complementary committee at the provincial level was also appointed.

The first National Food and Nutrition Plan (1977–1981) provided for a policy commitment to actions that would combat protein-energy malnutrition among mothers, pregnant women, children under 5 and school-aged children.

Lessons learned

Despite policy commitments to inter-sectoral and intra-sectoral collaboration, the programme was not fully implemented because of poor planning, the underfunding of multi-sectoral interventions and an over-emphasis on vertical planning. As a result, by 1980 surveys were finding that under-nutrition remained a serious problem among all groups and that 53 percent of children under 5 were suffering from protein-energy malnutrition. But in spite of evident failures, the programme did create significant awareness of nutritional problems among policymakers, the public and the private sector.

Modifying strategies

The National Health Development Plan 1982–1986 (NHDP) addressed these lessons, providing for a highly participatory community-based approach. There was commitment to community healthcare:

- 50,000 village health volunteers and 500,000 village health communicators were trained;
- there was at least one hospital in each district thus freeing up and engaging more fully health facilities, and community hospitals and sub-district health centres; and,
- the NHDP recognized that undernutrition was a manifestation of poverty and lack of knowledge, and so targeted interventions were developed for mothers, pregnant women, infants and children.

The plan also promoted quantifiable goals, including the eradication of iodine-deficiency goitre in nine endemic provinces.

Moving forward

The emphasis on community-level action contrasted with Thailand's traditional trickle-down development approach. The strategies employed to address undernutrition and improve the nutritional status of the population included nutrition surveillance, nutrition information, education and communication, production of nutritious foods in communities, supplementary food production and supplementary feeding programmes at the village level, school lunch programmes in 5,000 schools, food fortification, and training.

Results

Within nine years, the prevalence of protein-energy malnutrition in children under 5 was reduced from 51 percent to 21 percent; moderate to severe protein-energy malnutrition was almost eliminated; it fell from 15 percent to less than 1 percent. Food security was improved through yield-enhancing agricultural technologies and freshwater fisheries, livestock and other efforts. Primary healthcare reached 80 percent of the targeted villages. Village health communicators and volunteers were trained. Goal setting and accountability mechanisms with communities resulted in improvements in planning at all levels, integration of development activities and improved nutrition management.

Thailand's approach achieved excellent results in alleviating undernutrition. It took 10–15 years: 5 to 6 years were needed to create awareness and strong political commitment. During the subsequent period of implementation, political support was maintained, managerial and coordination structures were established, development objectives were identified and community participation was recognized as the cornerstone of the development process.

Contributed to the World Hunger Series *by Rosemary Fernholz and Channa Jayasekera, Duke University.*

Map 8 – National commitment to health

The boundaries and the designations used on this map do not imply any official endorsement or acceptance by the United Nations.
Map produced by WFP VAM.

Data source: WHO, 2007

Russian Federation
Kazakhstan
Mongolia
Georgia
Uzbekistan
Kyrgyzstan
Turkmenistan
Iraq
Islamic Rep. of Iran
Afghanistan
Jammu and Kashmir
China
Dem. People's Rep. of Korea
Rep. of Korea
Japan
Pakistan
Nepal
Saudi Arabia
Qatar
Arab Emirates
India
Oman
Yemen
Bangladesh
Myanmar
Viet Nam
Lao Peo. Dem.
Taiwan
Hong Kong (China)
Macao (China)
Thailand
Northern Mariana Islands (USA)
Philippines
Guam (USA)
Cambodia
Eritrea
Djibouti
Ethiopia
Somalia
Sri Lanka
Maldives
Brunei Darussalam
Palau
Federated States of Micronesia
Kiribati
Malaysia
Kenya
Singapore
Nauru
Tuvalu
Seychelles
Tanzania
Indonesia
Papua New Guinea
Solomon Is.
Wallis and Futune Islands (Fr.)
Timor-Leste
Christmas (Australia)
Comoros
Cocos (Keeling) Islands (Australia)
Madagascar
Mozambique
Vanuatu
Mauritius
Réunion (Fr.)
New Caledonia (Fr.)
Australia
New Zealand
wfp vam
vulnerability analysis and mapping

PART V Resource Compendium

Which countries are making the most rapid progress in addressing hunger and health issues?

The **Resource Compendium** offers the latest available data on the state of hunger and health in developing countries.

Overview – technical notes

This compendium is based on ten questions covering some of the main issues that policy-makers consider when addressing hunger and poor health. The following ten tables provide indicative responses to each question using 68 health and hunger indicators. The tables serve as a starting point for assessing how well a country is doing for each indicator.

Table structure

The title poses a question that the table helps to answer. The table header is divided into groups of indicators, followed by a definition for each indicator and the year in which it was collected.

A year span followed by an asterisk – for example, 1996–2006* – indicates that the data are for the most recent year in that period for which data are available. A year span followed by two asterisks – for example 1990–2006** – indicates progress made for each indicator from the baseline year to the latest year for which data are available.

A dash (–) indicates no data available.

Zero (0) means that the value is zero.

Each table is accompanied by definitions, calculations, methods and data sources.

Indicator terminology

Descriptions of the indicators have been kept as short as possible without changing the meaning of the indicator as provided by the source. This may cause certain inconsistencies in the tables, as different sources use different wordings.

Data sources and methods

The data in the compendium are derived from a number of sources. The main source was WHO, whose regional classifications (see below) were applied for the construction of all tables and graphs. Other sources include FAO, UNICEF, United Nations High Commission for Refugees (UNHCR) and the United States Department of Agriculture (USDA). Wherever possible, data are presented from original sources or from the institution mandated to collate national-level data. Unless otherwise indicated, data are from the most recent year for which reliable data are available from the indicated source, which may not be the same year for each data set.

Classification of countries

The tables present data for 166 developing countries and countries in transition, of which 82 are LIFDCs. These designations do not express a judgement about the development of a particular country. The term "country" does not imply political independence, but may refer to any territory for which authorities report separate statistics. Regional classifications follow the WHO system.

Definition of low-income food deficit countries

Low-income food-deficit countries, as defined by FAO, include all food-deficit/net cereal importing countries with a per capita income below the historical ceiling used by the World Bank to determine eligibility for International Development Association assistance and for 20-year International Bank for Reconstruction and Development terms. The designation LIFDC is applied to countries included in World Bank categories I and II. The historical ceiling of per capita GNI for 2003, based on the *World Bank Atlas* method, is US$1,465. In 2006, 82 countries were classified by FAO as LIFDCs.

The official list of LIFDCs in 2005 are:

Afghanistan, Albania, Angola, Armenia, Azerbaijan, Bangladesh, Belarus, Benin, Bhutan, Bosnia and Herzegovina, Burkina Faso, Burundi, Cambodia, Cameroon, Cape Verde, Central African Republic, Chad, China, the Comoros, the Republic of the Congo, the Democratic Republic of the Congo, Côte d'Ivoire, Djibouti, Ecuador, Egypt, Equatorial Guinea, Eritrea, Ethiopia, Gambia, Georgia, Ghana, Guinea, Guinea-Bissau, Haiti, Honduras, India, Indonesia, Iraq, Kenya, Kiribati, the People's Democratic Republic of Korea, Kyrgyzstan, the Lao People's Democratic Republic, Lesotho, Liberia, Madagascar, Malawi, Mali, Mauritania, Mongolia, Morocco, Mozambique, Nepal, Nicaragua, Niger, Nigeria, Pakistan, Papua New Guinea, the Philippines, Rwanda, São Tomé and Principe, Senegal, Sierra Leone, the Solomon Islands, Somalia, Sri Lanka, Sudan, Swaziland, Syrian Arab Republic, Tajikistan, the United Republic of Tanzania, Timor-Leste, Togo, Tonga, Turkmenistan, Tuvalu, Uganda, Uzbekistan, Vanuatu, Yemen, Zambia and Zimbabwe.

LIFDCs are highlighted in red type in the tables.

Regional breakdowns

Reference to the subgroups "sub-Saharan Africa" and "South Asia" follow the UNICEF classification for selected tables in this report. However, the Data Compendium follows the WHO breakdown.

Sub-Saharan Africa

Angola, Benin, Botswana, Burkina Faso, Burundi, Cameroon, Cape Verde, Central African Republic, Chad, the Comoros, the Republic of the Congo, the Democratic Republic of the Congo, Côte d'Ivoire, Equatorial Guinea, Eritrea, Ethiopia, Gabon, Gambia, Ghana, Guinea, Guinea-Bissau, Kenya, Lesotho, Liberia, Madagascar, Malawi, Mali, Mauritania, Mauritius, Mozambique, Namibia, Niger, Nigeria, Rwanda, São Tomé and Principe, Senegal, the Seychelles, Sierra Leone, Somalia, South Africa, Swaziland, the United Republic of Tanzania, Togo, Uganda, Zambia, Zimbabwe.

South Asia

Afghanistan, Bangladesh, Bhutan, India, the Maldives, Nepal, Pakistan, Sri Lanka.

Table 1 – What does a hungry world look like?

	Undernourishment							Food availability			
	Total population (millions)		Number of people undernourished (millions)		Proportion of undernourished in total population (%)		Change in prevalence	Dietary energy consumption (kcal/person/day)		Nutrition gap (thousand tons)	Distribution food gap (thousand tons)
	1990–92	2001–03	1990–92	2001–03	1990–92	2001–03	1992–2002	1990–92	2001–03		
AFRICAN REGION											
Algeria	25.6	31.3	1.3	1.5	5.0	5.0	0.9	2,920	3,040	0	0
Angola	9.6	13.2	5.6	5.0	58.0	38.0	0.7	1,780	2,070	0	68
Benin	4.8	6.6	1.0	0.9	20.0	14.0	0.7	2,330	2,530	0	0
Botswana	1.4	1.8	0.3	0.5	23.0	30.0	1.3	2,260	2,180	–	–
Burkina Faso	9.2	12.6	1.9	2.1	21.0	17.0	0.8	2,350	2,460	0	221
Burundi	5.7	6.6	2.7	4.5	48.0	67.0	1.4	1,900	1,640	502	583
Cameroon	12.0	15.7	4.0	4.0	33.0	25.0	0.8	2,120	2,270	0	66
Cape Verde	–	–	–	–	–	–	–	–	–	0	0
Central African Republic	3.0	3.8	1.5	1.7	50.0	45.0	0.9	1,860	1,940	88	231
Chad	6.0	8.3	3.5	2.7	58.0	33.0	0.6	1,780	2,160	393	530
Comoros	–	–	–	–	–	–	–	–	–	–	–
Congo, Democratic Rep. of the	38.8	51.3	12.2	37.0	31.0	72.0	2.3	2,170	1,610	4,260	4,709
Congo, Rep. of the	2.6	3.6	1.4	1.2	54.0	34.0	0.6	1,860	2,150	–	–
Côte d'Ivoire	12.9	16.4	2.3	2.2	18.0	14.0	0.8	2,470	2,630	0	141
Equatorial Guinea	–	–	–	–	–	–	–	–	–	–	–
Eritrea	3.2	4.0	2.2	2.9	68.0	73.0	1.1	1,550	1,520	429	456
Ethiopia	55.6	69.0	38.2	31.5	61.0	46.0	0.8	1,550	1,860	3,261	3,664
Gabon	1.0	1.3	0.1	0.1	10.0	5.0	0.5	2,450	2,670	–	–
Gambia	1.0	1.4	0.2	0.4	22.0	27.0	1.2	2,370	2,280	0	21
Ghana	15.7	20.5	5.8	2.4	37.0	12.0	0.3	2,080	2,650	0	34
Guinea	6.4	8.4	2.5	2.0	39.0	24.0	0.6	2,110	2,420	0	88
Guinea Bissau	–	–	–	–	–	–	–	–	–	28	58
Kenya	24.4	31.5	9.5	9.7	39.0	31.0	0.8	1,980	2,150	265	940
Lesotho	1.6	1.8	0.3	0.2	17.0	12.0	0.7	2,440	2,620	152	192
Liberia	2.1	3.2	0.7	1.6	34.0	49.0	1.4	2,210	1,940	0	64
Madagascar	12.3	16.9	4.3	6.5	35.0	38.0	1.1	2,080	2,040	0	412
Malawi	9.6	11.9	4.8	4.0	50.0	34.0	0.7	1,880	2,140	117	453
Mali	9.3	12.6	2.7	3.5	29.0	28.0	1	2,220	2,220	0	161
Mauritania	2.1	2.8	0.3	0.3	15.0	10.0	0.6	2,560	2,780	0	7
Mauritius	1.1	1.2	0.1	0.1	6.0	6.0	0.9	2,890	2,960	–	–
Mozambique	13.9	18.5	9.2	8.3	66.0	45.0	0.7	1,730	2,070	0	96
Namibia	1.5	2.0	0.5	0.4	34.0	23.0	0.7	2,070	2,260	–	–
Niger	7.9	11.5	3.2	3.7	41.0	32.0	0.8	2,020	2,160	37	564
Nigeria	88.7	120.9	11.8	11.5	13.0	9.0	0.7	2,540	2,700	0	0
Rwanda	6.4	8.2	2.8	3.0	43.0	36.0	0.8	1,950	2,070	0	26
Sao Tome and Principe	–	–	–	–	–	–	–	–	–	–	–
Senegal	7.5	9.9	1.8	2.2	23.0	23.0	1	2,280	2,310	0	105
Seychelles	–	–	–	–	–	–	–	–	–	–	–
Sierra Leone	4.1	4.8	1.9	2.4	46.0	50.0	1.1	1,990	1,930	199	451
South Africa	–	–	–	–	–	–	–	–	–	–	–
Swaziland	0.9	1.1	0.1	0.2	14.0	19.0	1.3	2,450	2,360	0	14
Tanzania, United Rep. of	27.0	36.3	9.9	16.1	37.0	44.0	1.2	2,050	1,960	706	1,028
Togo	3.5	4.8	1.2	1.2	33.0	25.0	0.7	2,150	2,320	227	281
Uganda	17.9	25.0	4.2	4.6	24.0	19.0	0.8	2,270	2,380	0	205
Zambia	8.4	10.7	4.0	5.1	48.0	47.0	1	1,930	1,930	428	663
Zimbabwe	10.7	12.8	4.8	5.7	45.0	45.0	1	1,980	2,010	685	862
REGION OF THE AMERICAS											
Antigua and Barbuda	–	–	–	–	–	–	–	–	–	–	–
Argentina	33.0	38.0	0.7	0.9	–	–	1.1	3,000	2,980	–	–
Bahamas	–	–	–	–	–	–	–	–	–	–	–

	Undernourishment							Food availability			
	Total population (millions)		Number of people undernourished (millions)		Proportion of undernourished in total population (%)		Change in prevalence	Dietary energy consumption (kcal/person/day)		Nutrition gap (thousand tons)	Distribution food gap (thousand tons)
	1990–92	2001–03	1990–92	2001–03	1990–92	2001–03	1992–2002	1990–92	2001–03		
Barbados	–	–	–	–	–	–	–	–	–	–	–
Belize	–	–	–	–	–	–	–	–	–	–	–
Bolivia	6.8	8.6	1.9	2.0	28.0	23.0	0.8	2,110	2,220	0	128
Brazil	151.2	176.3	18.5	14.4	12.0	8.0	0.7	2,810	3,060	–	–
Chile	13.3	15.6	1.1	0.6	8.0	4.0	0.5	2,610	2,860	–	–
Colombia	35.7	43.5	6.1	5.9	17.0	14.0	0.8	2,440	2,580	0	440
Costa Rica	3.2	4.1	0.2	0.2	6.0	4.0	0.8	2,720	2,850	–	–
Cuba	10.7	11.3	0.7	0.2	7.0	–	0.2	2,720	3,190	–	–
Dominica	–	–	–	–	–	–	–	–	–	–	–
Dominican Republic	7.2	8.6	1.9	2.3	27.0	27.0	1	2,260	2,290	0	26
Ecuador	10.5	12.8	0.9	0.6	8.0	5.0	0.6	2,510	2,710	0	179
El Salvador	5.2	6.4	0.6	0.7	12.0	11.0	0.9	2,490	2,560	0	73
Grenada	–	–	–	–	–	–	–	–	–	–	–
Guatemala	9.0	12.0	1.4	2.8	16.0	23.0	1.5	2,350	2,210	0	289
Guyana	0.7	0.8	0.2	0.1	21.0	9.0	0.4	2,350	2,730	–	–
Haiti	7.0	8.2	4.6	3.8	65.0	47.0	0.7	1,780	2,090	0	221
Honduras	5.0	6.8	1.1	1.5	23.0	22.0	1	2,310	2,360	281	415
Jamaica	2.4	2.6	0.3	0.3	14.0	10.0	0.7	2,500	2,680	0	0
Mexico	84.8	102.0	4.6	5.1	5.0	5.0	0.9	3,100	3,180	–	–
Nicaragua	3.9	5.3	1.2	1.5	30.0	27.0	0.9	2,220	2,290	0	134
Panama	2.5	3.1	0.5	0.8	21.0	25.0	1.2	2,320	2,260	–	–
Paraguay	4.3	5.7	0.8	0.8	18.0	15.0	0.8	2,400	2,530	–	–
Peru	22.2	26.8	9.3	3.3	42.0	12.0	0.3	1,960	2,570	0	226
Saint Kitts and Nevis	–	–	–	–	–	–	–	–	–	–	–
Saint Lucia	–	–	–	–	–	–	–	–	–	–	–
Saint Vincent and the Grenadines	–	–	–	–	–	–	–	–	–	–	–
Suriname	0.4	0.4	0.1	0.0	13.0	10.0	0.7	2,530	2,660	–	–
Trinidad and Tobago	1.2	1.3	0.2	0.1	13.0	11.0	0.8	2,630	2,760	–	–
Uruguay	3.1	3.4	0.2	0.1	7.0	3.0	0.5	2,660	2,850	–	–
Venezuela, Bolivarian Rep. of	20.0	25.2	2.3	4.5	11.0	18.0	1.6	2,460	2,350	–	–
SOUTH EAST ASIA REGION											
Bangladesh	112.1	143.8	39.2	43.1	35.0	30.0	0.9	2,070	2,200	0	231
Bhutan	–	–	–	–	–	–	–	–	–	–	–
India	863.3	1,049.5	214.8	212.0	25.0	20.0	0.8	2,370	2,440	0	1,152
Indonesia	185.2	217.1	16.4	13.8	9.0	6.0	0.7	2,700	2,880	0	0
Korea, Democratic People's Rep. of	20.3	22.5	3.6	7.9	18.0	35.0	2	2,470	2,150	7	362
Maldives	–	–	–	–	–	–	–	–	–	–	–
Myanmar	41.2	48.8	4.0	2.7	10.0	5.0	0.6	2,630	2,900	–	–
Nepal	19.1	24.6	3.9	4.1	20.0	17.0	0.8	2,340	2,450	0	223
Sri Lanka	17.0	18.9	4.8	4.1	28.0	22.0	0.8	2,230	2,390	0	0
Thailand	55.1	62.2	16.8	13.4	30.0	21.0	0.7	2,200	2,410	–	–
Timor-Leste	–	–	–	–	–	–	–	–	–	–	–
EUROPEAN REGION											
Albania	3.2	3.1	0.2	0.2	5.0	6.0	1.1	2,870	2,860	–	–
Armenia	3.4	3.1	1.8	0.9	52.0	29.0	0.6	1,960	2,260	–	–
Azerbaijan	7.7	8.3	2.6	0.8	34.0	10.0	0.3	2,140	2,620	0	0
Belarus	10.3	9.9	0.1	0.3	–	3.0	2.8	3,190	2,960	–	–
Bosnia and Herzegovina	3.6	4.1	0.3	0.4	9.0	9.0	1	2,690	2,710	–	–
Bulgaria	8.5	8.0	0.7	0.7	8.0	9.0	1.2	2,900	2,850	–	–

	Undernourishment							Food availability			
	Total population (millions)		Number of people undernourished (millions)		Proportion of undernourished in total population (%)		Change in prevalence	Dietary energy consumption (kcal/person/day)		Nutrition gap (thousand tons)	Distribution food gap (thousand tons)
	1990–92	2001–03	1990–92	2001–03	1990–92	2001–03	1992–2002	1990–92	2001–03		
Croatia	4.5	4.4	0.7	0.3	16.0	7.0	0.4	2,520	2,770	–	–
Cyprus	–	–	–	–	–	–	–	–	–	–	–
Czech Republic	10.3	10.2	0.2	0.1	–	–	0.6	3,080	3,240	–	–
Estonia	1.5	1.3	0.1	0.0	9.0	3.0	0.3	2,760	3,160	–	–
Georgia	5.4	5.2	2.4	0.7	44.0	13.0	0.3	2,050	2,520	0	0
Hungary	10.2	9.9	0.1	0.0	–	–	0.6	3,340	3,500	–	–
Israel	–	–	–	–	–	–	–			–	–
Kazakhstan	16.7	15.5	0.2	1.2	–	8.0	7.8	3,280	2,710	0	0
Kyrgyzstan	4.5	5.1	1.0	0.2	21.0	4.0	0.2	2,400	3,050	0	0
Latvia	2.5	2.3	0.1	0.1	3.0	3.0	0.9	2,960	3,020	–	–
Lithuania	3.6	3.5	0.2	0.0	4.0	–	0.2	2,870	3,370	–	–
Macedonia, The Former Yugoslav Rep. of	2.0	2.0	0.3	0.1	15.0	7.0	0.4	2,520	2,800	–	–
Malta	–	–	–	–	–	–	–	–	–	–	–
Moldova, Rep. of	4.4	4.3	0.2	0.5	5.0	11.0	1.9	2,930	2,730	–	–
Poland	38.5	38.6	0.3	0.3	–	–	1	3,340	3,370	–	–
Romania	22.8	22.4	0.3	0.1	–	–	0.3	3,210	3,520	–	–
Russian Federation	148.4	144.1	6.4	4.1	4.0	3.0	0.7	2,930	3,080	–	–
Serbia and Montenegro	10.5	10.5	0.5	1.1	5.0	10.0	2.2	2,910	2,670	–	–
Slovakia	5.3	5.4	0.2	0.3	4.0	6.0	1.6	2,920	2,830	–	–
Slovenia	2.0	2.0	0.1	0.1	3.0	3.0	0.9	2,950	2,970	–	–
Tajikistan	5.7	6.2	1.2	3.8	22.0	61.0	2.8	2,310	1,840	108	169
Turkey	58.7	70.3	1.0	2.0	–	3.0	1.7	3,490	3,340	–	–
Turkmenistan	4.1	4.8	0.5	0.4	12.0	8.0	0.7	2,550	2,750	0	0
Ukraine	51.7	48.9	1.2	1.2	–	3.0	1.1	3,040	3,030	–	–
Uzbekistan	22.3	25.7	1.7	6.7	8.0	26.0	3.4	2,660	2,270	0	0
EASTERN MEDITERRANEAN REGION											
Afghanistan	–	–	–	–	–	–	–	–	–	220	833
Bahrain	–	–	–	–	–	–	–	–	–	–	–
Djibouti	–	–	–	–	–	–	–	–	–	–	–
Egypt	57.0	70.5	2.5	2.4	4.0	3.0	0.8	3,200	3,350	0	0
Iran, Islamic Rep. of	58.0	68.1	2.1	2.7	4.0	4.0	1.1	2,980	3,090	–	–
Iraq	–	–	–	–	–	–	–	–	–	–	–
Jordan	3.4	5.3	0.1	0.4	4.0	7.0	1.9	2,820	2,680	–	–
Kuwait	2.1	2.4	0.5	0.1	24.0	5.0	0.2	2,340	3,060	–	–
Lebanon	2.8	3.6	0.1	0.1	–	3.0	1.2	3,160	3,170	–	–
Libyan Arab Jamahiriya	4.4	5.4	0.0	0.0	–	–	1.1	3,270	3,330	–	–
Morocco	25.0	30.1	1.5	1.9	6.0	6.0	1.1	3,030	3,070	0	0
Oman	–	–	–	–	–	–	–	–	–	–	–
Pakistan	113.7	149.9	27.8	35.2	24.0	23.0	1	2,300	2,340	0	220
Qatar	–	–	–	–	–	–	–	–	–	–	–
Saudi Arabia	17.1	23.5	0.7	0.9	4.0	4.0	0.9	2,770	2,820	–	–
Somalia	–	–	–	–	–	–	–	–	–	550	587
Sudan	25.5	32.9	7.9	8.8	31.0	27.0	–	2,170	2,260	31	592
Syrian Arab Republic	13.1	17.4	0.7	0.6	5.0	4.0	0.7	2,830	3,060	–	–
Tunisia	8.4	9.7	0.1	0.1	–	–	1	3,150	3,250	0	0
United Arab Emirates	2.1	2.9	0.1	0.1	4.0	–	0.4	2,930	3,220	–	–
Yemen	12.5	19.3	4.2	7.1	34.0	37.0	1.1	2,040	2,020	–	–

	Undernourishment							Food availability			
	Total population (millions)		Number of people undernourished (millions)		Proportion of undernourished in total population (%)		Change in prevalence	Dietary energy consumption (kcal/person/day)		Nutrition gap (thousand tons)	Distribution food gap (thousand tons)
	1990–92	2001–03	1990–92	2001–03	1990–92	2001–03	1992–2002	1990–92	2001–03		
WESTERN PACIFIC REGION											
Brunei Darussalam	–	–	–	–	–	–	–	–	–	–	–
Cambodia	10.1	13.8	4.4	4.6	43.0	33.0	0.8	1,860	2,060	–	–
China	1,175.7	1,302.2	193.6	150.0	16.0	12.0	0.7	2,710	2,940	–	–
Cook Islands	–	–	–	–	–	–	–	–	–	–	–
Fiji	–	–	–	–	–	–	–	–	–	–	–
Kiribati	–	–	–	–	–	–	–	–	–	–	–
Korea, Rep. of	43.3	47.4	0.8	0.8	–	–	0.9	3,000	3,040	–	–
Lao People's Democratic Rep.	4.2	5.5	1.2	1.2	29.0	21.0	0.7	2,110	2,320	–	–
Malaysia	18.3	24.0	0.5	0.6	3.0	3.0	1	2,830	2,870	–	–
Marshall Islands	–	–	–	–	–	–	–	–	–	–	–
Micronesia, Federated States of	–	–	–	–	–	–	–	–	–	–	–
Mongolia	2.3	2.6	0.8	0.7	34.0	28.0	0.8	2,060	2,250	–	–
Nauru	–	–	–	–	–	–	–	–	–	–	–
Niue	–	–	–	–	–	–	–	–	–	–	–
Palau	–	–	–	–	–	–	–	–	–	–	–
Papua New Guinea	–	–	–	–	–	–	–	–	–	–	–
Philippines	62.5	78.6	16.2	15.2	26.0	19.0	0.7	2,260	2,450	0	136
Samoa	–	–	–	–	–	–	–	–	–	–	–
Singapore	–	–	–	–	–	–	–	–	–	–	–
Solomon Islands	–	–	–	–	–	–	–	–	–	–	–
Tonga	–	–	–	–	–	–	–	–	–	–	–
Tuvalu	–	–	–	–	–	–	–	–	–	–	–
Vanuatu	–	–	–	–	–	–	–	–	–	–	–
Viet Nam	67.5	80.3	20.6	13.8	31.0	17.0	0.6	2,180	2,580	0	0

Indicator definitions

Change in prevalence: Division of the current value (2000–2002) by the baseline value (1990–1992). A value 0–1 indicates a reduction in prevalence; a value above 1 indicates that prevalence has increased since 1990–1992.

Dietary energy consumption: Amount of food available in the country in kilocalories per day divided by the total population and expressed as daily availability.

Nutrition gap: Amount of food needed to raise food consumption of an entire population to the minimum nutritional requirement, disregarding different income groups.

Distribution food gap: Amount of food needed to raise consumption in each income group to the nutritional target.

Sources

- **Population, undernourishment and food consumption:** Food and Agriculture Organization of the United Nations. 2006. *The State of Food Insecurity in the World 2006*. Rome.
- **Food availability:** United States Department of Agriculture. 2005. *Food Security Assessment Report GFA 17*. Washington DC.

Table 2 – How many people suffer from hunger throughout the life cycle?

	Prevalence of under 5 stunting (%)	Prevalence of under 5 underweight (%)		Prevalence of under 5 wasting (%)	Newborns with low birthweight (%)	Proportion of women with low body mass index (%)
	1997–2006*	1990	1997–2006*	1995–2005*	2002	1992–2002*
AFRICAN REGION						
Algeria	21.6	–	10.2	8	7	–
Angola	50.8	40.6	27.5	6	12	–
Benin	39.1	–	21.5	8	16	10.5
Botswana	29.1	–	10.7	5	10	–
Burkina Faso	43.1	32.7	35.2	19	19	13.2
Burundi	63.1	32.7	38.9	8	16	–
Cameroon	35.4	15.1	15.1	5	11	7.9
Cape Verde	–	–	–	–	13	–
Central African Republic	44.6	27.3	21.8	9	14	15.3
Chad	44.8	38.8	33.9	14	17	21
Comoros	46.9	18.5	25	8	25	10.3
Congo, Democratic Rep. of the	44.4	34.4	33.6	13	12	–
Congo, Rep. of the	31.2	23.9	11.8	7	–	–
Côte d'Ivoire	31.5	23.8	18.2	7	17	7.4
Equatorial Guinea	42.6	–	15.7	–	13	–
Eritrea	43.7	41	34.5	13	21	40.6
Ethiopia	50.7	46.2	34.6	11	15	26
Gabon	26.3	–	8.8	3	14	6.6
Gambia	24.1	26.2	15.4	8	17	–
Ghana	35.6	27.3	18.8	7	11	11.2
Guinea	39.3	–	22.5	9	12	11.9
Guinea Bissau	36.1	–	21.9	10	22	–
Kenya	35.8	22.6	16.5	6	11	11.9
Lesotho	53	15.8	15	4	14	–
Liberia	45.3	–	22.8	6	–	–
Madagascar	52.8	40.9	36.8	13	14	20.6
Malawi	52.5	23.9	18.4	5	16	6.5
Mali	42.7	–	30.1	11	23	–
Mauritania	39.5	47.6	30.4	13	–	8.6
Mauritius	–	23.9	–	14	13	–
Mozambique	47	27	21.2	4	14	10.9
Namibia	29.5	–	20.3	9	14	13.8
Niger	54.2	42.6	43.6	14	17	20.7
Nigeria	43	35.5	27.2	9	14	16.2
Rwanda	48.3	29.4	20.3	4	9	5.9
Sao Tome and Principe	35.2	–	10.1	4	–	–
Senegal	20.1	21.6	14.5	8	18	–
Seychelles	–	5.7	–	–	–	–
Sierra Leone	38.4	–	24.7	10	–	–
South Africa	30.9	–	9.6	3	15	–
Swaziland	36.6	–	9.1	1	9	–
Tanzania, United Rep. of	44.4	28.9	16.7	3	13	–
Togo	29.8	–	23.2	12	15	10.9
Uganda	44.8	23	19	4	12	9.4
Zambia	52.5	20.5	23.3	6	12	13
Zimbabwe	33.7	11.5	11.5	5	11	4.5
REGION OF THE AMERICAS						
Antigua and Barbuda	–	–	–	–	8	–
Argentina	8.2	1.9	2.3	1	7	–
Bahamas	–	–	–	–	7	–

	Prevalence of under 5 stunting (%)	Prevalence of under 5 underweight (%)		Prevalence of under 5 wasting (%)	Newborns with low birthweight (%)	Proportion of women with low body mass index (%)
	1997–2006*	1990	1997–2006*	1995–2005*	2002	1992–2002*
Barbados	–	–	–	–	10	–
Belize	–	–	–	–	6	–
Bolivia	32.5	11.3	5.9	1	9	0.9
Brazil	–	7	3.7	2	10	6.2
Chile	2.7	2.5	0.8	0	5	–
Colombia	16.2	10.1	5.1	1	9	3.1
Costa Rica	–	2.8	–	2	7	–
Cuba	9.6	–	4.3	2	6	–
Dominica	–	–	–	–	10	–
Dominican Republic	11.7	10.3	4.2	2	11	–
Ecuador	29	16.5	6.2	–	16	–
El Salvador	24.6	15.2	6.1	1	13	–
Grenada	–	–	–	–	9	–
Guatemala	54.3	33.2	17.7	2	13	2
Guyana	13.8	18.3	11.9	11	12	–
Haiti	28.3	26.8	13.9	5	21	–
Honduras	29.9	18	8.6	1	14	–
Jamaica	4.5	4.6	3.1	4	9	–
Mexico	15.5	14.2	3.4	2	9	–
Nicaragua	25.2	11	7.8	2	12	3.8
Panama	21.5	15.8	6.3	1	10	–
Paraguay	–	–	–	1	9	–
Peru	31.3	10.7	5.2	1	11	0.7
Saint Kitts and Nevis	–	–	–	–	9	–
Saint Lucia	–	–	–	–	8	–
Saint Vincent and the Grenadines	–	–	–	–	10	–
Suriname	14.5	–	11.4	7	13	–
Trinidad and Tobago	5.3	6.7	4.4	4	23	–
Uruguay	13.9	7.4	6	1	8	–
Venezuela, Bolivarian Rep. of	16.7	7.7	4.8	4	7	–
SOUTH EAST ASIA REGION						
Bangladesh	50.5	65.8	42.7	13	30	45.4
Bhutan	47.7	37.9	14.1	3	15	–
India	51	56.1	44.4	16	30	41.2
Indonesia	28.6	35.5	19.7	–	9	–
Korea, Democratic People's Rep. of	44.7	–	17.8	7	7	–
Maldives	31.9	39	25.7	13	22	–
Myanmar	40.6	32.4	29.6	9	15	–
Nepal	57.1	48.5	43	10	21	26.6
Sri Lanka	18.4	37.3	22.8	14	22	–
Thailand	15.5	25.3	7.3	5	9	–
Timor-Leste	55.7	–	40.6	12	10	–
EUROPEAN REGION						
Albania	39.2	8.1	17	11	3	–
Armenia	18.2	3.3	4.2	5	7	5
Azerbaijan	24.1	10.1	14	2	11	–
Belarus	–	–	–	–	5	–
Bosnia and Herzegovina	12.1	–	4.2	6	4	–
Bulgaria	8.8	–	1.6	–	10	–
Croatia	–	–	–	1	6	–

	Prevalence of under 5 stunting (%)	Prevalence of under 5 underweight (%)		Prevalence of under 5 wasting (%)	Newborns with low birthweight (%)	Proportion of women with low body mass index (%)
	1997–2006*	1990	1997–2006*	1995–2005*	2002	1992–2002*
Cyprus	–	–	–	–	7	–
Czech Republic	2.6	–	2.1	–	4	–
Estonia	–	–	–	–	6	–
Georgia	15.2	–	3.4	2	9	–
Hungary	–	–	–	–	8	–
Israel	–	–	–	–	8	–
Kazakhstan	13.9	–	3.8	2	7	9.8
Kyrgyzstan	32.6	–	8.2	3	5	6.2
Latvia	–	–	–	–	4	–
Lithuania	–	–	–	–	6	–
Macedonia, The Former Yugoslav Rep. of	1.2	–	1.2	4	5	–
Malta	–	–	–	–	0	–
Moldova, Rep. of	11.3	–	3.2	4	5	–
Poland	–	–	–	–	6	–
Romania	12.8	–	3.5	2	9	–
Russian Federation	–	–	–	4	6	–
Serbia and Montenegro	9.8	–	2.2	–	4	–
Slovakia	–	–	–	–	7	–
Slovenia	–	–	–	–	6	–
Tajikistan	42	–	–	5	15	–
Turkey	19.1	10.4	7	1	16	2.6
Turkmenistan	27.7	–	10	6	6	10.1
Ukraine	5.6	–	4.1	0	5	–
Uzbekistan	26.2	–	6.2	7	7	9.8
EASTERN MEDITERRANEAN REGION						
Afghanistan	53.6	–	46.2	7	–	–
Bahrain	–	7.2	–	5	8	–
Djibouti	28.6	22.9	23.9	18	–	–
Egypt	23.8	10.4	5.4	4	12	0.6
Iran, Islamic Rep. of	19.7	–	9.1	5	7	–
Iraq	28.3	11.9	12.9	8	15	–
Jordan	12	6.4	3.6	2	10	2.3
Kuwait	6.7	10.5	1.9	11	7	–
Lebanon	5.8	–	4.3	5	6	–
Libyan Arab Jamahiriya	–	–	–	–	7	–
Morocco	23.1	9.5	9.9	9	11	3.9
Oman	15.9	24.3	13.1	7	8	–
Pakistan	41.5	40.2	31.3	13	19	–
Qatar	–	–	–	2	10	–
Saudi Arabia	–	–	–	11	11	–
Somalia	29	–	23	17	–	–
Sudan	47.6	33.9	38.4	16	31	–
Syrian Arab Republic	24.1	12.1	9.1	4	6	–
Tunisia	16	9	4.4	2	7	–
United Arab Emirates	–	–	–	15	15	–
Yemen	59.8	30	42.7	12	32	25.2
WESTERN PACIFIC REGION						
Brunei Darussalam	–	–	–	–	10	–
Cambodia	49.2	47.4	39.5	15	11	21.2
China	18.6	17.4	6.1	–	6	–

	Prevalence of under 5 stunting (%)	Prevalence of under 5 underweight (%)		Prevalence of under 5 wasting (%)	Newborns with low birthweight (%)	Proportion of women with low body mass index (%)
	1997–2006*	1990	1997–2006*	1995–2005*	2002	1992–2002*
Cook Islands	–	–	–	–	3	–
Fiji	–	–	–	–	10	–
Kiribati	–	–	–	–	5	–
Lao People's Democratic Rep.	48.2	44	36.4	15	14	–
Malaysia	20	25	16.2	–	10	–
Marshall Islands	–	–	–	–	12	–
Micronesia, Federated States of	–	–	–	–	18	–
Mongolia	23.5	12.3	4.8	3	8	–
Nauru	–	–	–	–	–	–
Niue	–	–	–	–	–	–
Palau	–	–	–	–	9	–
Papua New Guinea	43.9	–	18.1	–	11	–
Philippines	33.8	33.5	20.7	6	20	–
Korea, Rep. of	–	–	–	–	4	–
Samoa	8.7	–	2.1	–	4	–
Singapore	4.4	14.4	3.3	2	8	–
Solomon Islands	–	–	–	–	13	–
Tonga	–	–	–	–	–	–
Tuvalu	–	–	–	–	5	–
Vanuatu	–	–	–	–	6	–
Viet Nam	43.4	45	26.7	8	9	–

Indicator definitions

Prevalence of stunting in children under 5: Prevalence of children below –2 standard deviations from the median height-for-age of the reference population.

Prevalence of underweight in children under 5: Includes moderate underweight – proportion of children below –2 standard deviations from median weight-for-age of the reference population – and severe underweight – proportion of children below –3 standard deviations from median weight-for-age of the reference population.

Prevalence of wasting in children under 5: Prevalence of children below –2 standard deviations from the median weight-for-height of the reference population.

Newborns with low birthweight: Percentage of live births weighing less than 2.5 kg or 5.5 pounds.

Low body mass index: Body mass index (BMI) determined by dividing weight in kilograms by the square of height in metres. A value of less than 18.5 indicates underweight.

Sources

- **Underweight, stunting, low birthweight:** World Health Organization. 2007. *World Health Statistics 2007*. Geneva.
- **Wasting:** United Nations Children's Fund. 2006. *The State of the World's Children 2007 – Women and Children. The Double Dividend of Gender Equality*. New York.
- **BMI:** Standing Committee on Nutrition. 2004. *5th Report on the World Nutrition Situation*. Geneva.

Note

A year span followed by an asterisk indicates that the data are from the most recent year in that period for which data are available.

Table 3 – What does a world with poor health look like?

	Life expectancy at birth (years)		Healthy life expectancy at birth (years)		Access to health services				Access to improved drinking-water resources (%)		Access to improved sanitation facilities (%)	
					Health care workers		Basic health services					
	Male 2005	Female 2005	Male 2002	Female 2002	Physicians (per 1,000 pop.)	Nurses (per 1,000 pop.)	Immunization rate (DPT3)	ANC coverage at least 4 visits	Urban pop. 2004 (%)	Rural pop. 2004 (%)	Urban pop. 2004 (%)	Rural pop. 2004 (%)
AFRICAN REGION												
Algeria	70	72	60	62	1.13	2.21	88	–	88	80	99	82
Angola	39	41	32	35	0.08	1.15	47	–	75	40	56	16
Benin	52	53	43	45	0.04	0.84	93	61	78	57	59	11
Botswana	42	41	36	35	0.4	2.65	97	97	100	90	57	25
Burkina Faso	48	49	35	36	0.06	0.41	96	18	94	54	42	6
Burundi	46	48	33	37	0.03	0.19	74	79	92	77	47	35
Cameroon	50	51	41	42	0.19	1.6	80	52	86	44	58	43
Cape Verde	67	72	59	63	0.49	0.87	73	99	86	73	61	19
Central African Republic	42	42	37	38	0.08	0.3	40	39	93	61	47	12
Chad	46	48	40	42	0.04	0.27	20	13	41	43	24	4
Comoros	62	67	54	55	0.15	0.74	80	53	92	82	41	29
Congo, Democratic Rep. of the	44	48	35	39	0.11	0.53	73	–	82	29	42	25
Congo, Rep. of the	54	55	45	47	0.2	0.96	65	–	84	27	28	25
Côte d'Ivoire	42	47	38	41	0.12	0.6	56	35	97	74	46	29
Equatorial Guinea	45	47	45	46	0.3	0.45	33	37	45	42	60	46
Eritrea	59	63	49	51	0.05	0.58	83	49	74	57	32	3
Ethiopia	50	53	41	42	0.03	0.21	69	10	81	11	44	7
Gabon	54	57	50	53	0.29	5.16	38	63	95	47	37	30
Gambia	53	57	48	51	0.11	1.21	88	–	95	77	72	46
Ghana	56	58	49	50	0.15	0.92	84	69	88	64	27	11
Guinea	53	55	44	46	0.11	0.55	69	48	78	35	31	11
Guinea Bissau	46	48	40	41	0.12	0.67	80	62	79	49	57	23
Kenya	51	51	44	45	0.14	1.14	76	52	83	46	46	41
Lesotho	42	41	30	33	0.05	0.62	83	88	92	76	61	32
Liberia	41	44	34	37	0.03	0.18	87	84	72	52	49	7
Madagascar	56	60	47	50	0.29	0.32	61	38	77	35	48	26
Malawi	47	46	35	35	0.02	0.59	93	55	98	68	62	61
Mali	45	47	37	38	0.08	0.49	85	30	78	36	59	39
Mauritania	55	60	43	46	0.11	0.64	71	16	59	44	49	8
Mauritius	69	76	60	65	1.06	3.69	97	–	100	100	95	94
Mozambique	46	45	36	38	0.03	0.21	72	41	72	26	53	19
Namibia	52	52	43	44	0.3	3.06	86	69	98	81	50	13
Niger	42	41	36	35	0.03	0.22	89	11	80	36	43	4
Nigeria	47	48	41	42	0.28	1.7	25	47	67	31	53	36
Rwanda	44	47	36	40	0.05	0.42	95	10	92	69	56	38
Sao Tome and Principe	57	60	54	55	0.49	1.55	97	–	89	73	32	20
Senegal	54	57	47	49	0.06	0.32	84	64	92	60	79	34
Seychelles	68	77	57	65	1.51	7.93	99	–	100	75	–	100
Sierra Leone	37	40	27	30	0.03	0.36	64	68	75	46	53	30
South Africa	50	52	43	45	0.77	4.08	94	72	99	73	79	46
Swaziland	38	37	33	35	0.16	6.3	71	–	87	54	59	44
Tanzania, United Rep. of	48	50	40	41	0.02	0.37	90	69	87	56	54	41
Togo	52	56	44	46	0.04	0.43	82	46	85	49	53	43
Uganda	48	51	42	44	0.08	0.61	84	40	80	36	71	15
Zambia	40	40	35	35	0.12	1.74	80	71	90	40	59	52
Zimbabwe	43	42	34	33	0.16	0.72	–	64	98	72	63	47

	Life expectancy at birth (years)		Healthy life expectancy at birth (years)		Access to health services				Access to improved drinking-water resources (%)		Access to improved sanitation facilities (%)	
					Health care workers		Basic health services					
	Male 2005	Female 2005	Male 2002	Female 2002	Physicians (per 1,000 pop.)	Nurses (per 1,000 pop.)	Immunization rate (DPT3)	ANC coverage at least 4 visits	Urban pop. 2004 (%)	Rural pop. 2004 (%)	Urban pop. 2004 (%)	Rural pop. 2004 (%)
REGION OF THE AMERICAS												
Antigua and Barbuda	70	75	60	64	0.17	3.28	99	82	95	89	98	94
Argentina	72	78	62	68	3.01	0.8	92	95	98	80	92	83
Bahamas	70	76	61	66	1.05	4.47	93	–	98	86	100	100
Barbados	71	78	63	68	1.21	3.7	92	–	100	100	99	100
Belize	67	74	58	62	1.05	1.26	96	96	100	82	71	25
Bolivia	63	67	54	55	1.22	3.19	81	69	95	68	60	22
Brazil	68	75	57	62	1.15	3.84	96	76	96	57	83	37
Chile	74	81	65	70	1.09	0.63	91	95	100	58	95	62
Colombia	71	78	58	66	1.35	0.55	87	79	99	71	96	54
Costa Rica	75	80	65	69	1.32	0.92	91	70	100	92	89	97
Cuba	75	79	67	70	5.91	7.44	99	100	95	78	99	95
Dominica	72	76	62	66	0.5	4.17	98	100	100	90	86	75
Dominican Republic	65	72	57	62	1.88	1.84	77	93	97	91	81	73
Ecuador	70	75	60	64	1.48	1.57	94	–	97	89	94	82
El Salvador	69	74	57	62	1.24	0.8	89	76	94	70	77	39
Grenada	66	70	58	60	0.5	3.7	99	98	97	93	96	97
Guatemala	65	71	55	60	0.9	4.05	81	68	99	92	90	82
Guyana	63	64	53	57	0.48	2.29	93	–	83	83	86	60
Haiti	53	56	43	44	0.25	0.11	43	42	52	56	57	14
Honduras	65	70	56	61	0.57	1.29	91	84	95	81	87	54
Jamaica	70	74	64	66	0.85	1.65	88	99	98	88	91	69
Mexico	72	77	63	68	1.98	0.9	98	86	100	87	91	41
Nicaragua	68	73	60	63	0.37	1.07	86	72	90	63	56	34
Panama	74	78	64	68	1.5	1.54	85	72	99	79	89	51
Paraguay	70	76	60	64	1.11	1.69	75	89	99	68	94	61
Peru	70	74	60	62	1.17	0.67	84	69	89	65	74	32
Saint Kitts and Nevis	69	72	60	63	1.19	5.02	99	100	99	99	96	96
Saint Lucia	72	78	61	64	5.17	2.28	95	100	98	98	89	89
Saint Vincent and the Grenadines	66	74	60	62	0.87	2.38	99	92	–	93	–	96
Suriname	66	71	57	61	0.45	1.62	83	91	98	73	99	76
Trinidad and Tobago	67	74	60	64	0.79	2.87	95	98	92	88	100	100
Uruguay	71	79	63	69	3.65	0.85	96	94	100	100	100	99
Venezuela, Bolivarian Rep. of	72	78	62	67	1.94	–	87	90	85	70	71	48
SOUTH EAST ASIA REGION												
Bangladesh	62	63	55	53	0.26	0.14	88	11	82	72	51	35
Bhutan	62	65	53	53	0.05	0.14	95	–	86	60	65	70
India	62	64	53	54	0.6	0.8	59	30	95	83	59	22
Indonesia	66	69	57	59	0.13	0.62	70	81	87	69	73	40
Korea, Democratic People's Rep. of	65	68	58	60	3.29	3.85	79	–	79	43	67	20
Maldives	67	69	59	57	0.92	2.7	98	81	98	76	100	42
Myanmar	56	62	50	53	0.36	0.38	73	76	80	77	88	72
Nepal	61	61	52	51	0.21	0.22	75	15	96	89	62	30
Sri Lanka	68	75	59	64	0.55	1.58	99	98	98	74	98	89
Thailand	67	73	58	62	0.37	2.82	98	86	98	100	98	99
Timor-Leste	63	68	48	52	0.1	1.79	55	–	77	56	66	33

	Life expectancy at birth (years)		Healthy life expectancy at birth (years)		Access to health services				Access to improved drinking-water resources (%)		Access to improved sanitation facilities (%)	
					Health care workers		Basic health services					
	Male 2005	Female 2005	Male 2002	Female 2002	Physicians (per 1,000 pop.)	Nurses (per 1,000 pop.)	Immunization rate (DPT3)	ANC coverage at least 4 visits	Urban pop. 2004 (%)	Rural pop. 2004 (%)	Urban pop. 2004 (%)	Rural pop. 2004 (%)
EUROPEAN REGION												
Albania	69	73	59	63	1.31	3.62	98	42	99	94	99	84
Armenia	65	72	59	63	3.59	4.35	90	65	99	80	96	61
Azerbaijan	64	67	56	59	3.55	7.11	93	–	95	59	73	36
Belarus	63	75	57	65	4.55	11.63	99	–	100	100	93	61
Bosnia and Herzegovina	70	77	62	66	1.34	4.13	93	–	99	96	99	92
Bulgaria	69	76	63	67	3.56	3.75	96	–	100	97	100	96
Croatia	72	79	64	69	2.44	5.05	96	–	100	100	100	100
Cyprus	77	82	67	68	2.34	3.76	98	–	100	100	100	100
Czech Republic	73	79	66	71	3.51	9.71	97	–	100	100	99	97
Estonia	67	78	59	69	4.48	8.5	96	–	100	99	97	96
Georgia	68	75	62	67	4.09	3.47	84	–	96	67	96	91
Hungary	69	77	62	68	3.33	8.85	99	–	100	98	100	85
Israel	78	82	70	72	3.82	6.26	95	–	100	100	100	–
Kazakhstan	58	69	53	59	3.54	6.01	98	71	97	73	87	52
Kyrgyzstan	61	68	52	58	2.51	6.14	98	81	98	66	75	51
Latvia	65	76	58	68	3.01	5.27	99	–	100	96	82	71
Lithuania	65	77	59	68	3.97	7.62	94	–	–	–	–	–
Macedonia, The Former Yugoslav Rep. of	71	76	62	65	2.19	5.19	97	–	–	–	–	–
Malta	77	81	70	73	3.18	5.83	92	–	100	100	100	–
Moldova, Rep. of	65	72	57	62	2.64	6.06	98	–	97	88	86	52
Poland	71	79	63	68	2.47	4.9	99	–	–	–	–	–
Romania	68	76	61	65	1.9	3.89	97	–	91	16	89	–
Russian Federation	59	72	53	64	4.25	8.05	98	–	100	88	93	70
Serbia and Montenegro	70	76	63	69	2.06	4.64	–	–	99	86	97	77
Slovakia	70	78	67	72	3.18	6.77	99	–	100	99	100	98
Slovenia	74	81	70	75	2.25	7.21	96	–	–	–	–	–
Tajikistan	64	66	63	65	2.03	4.58	81	–	92	48	70	45
Turkey	69	74	61	63	1.35	1.7	90	42	98	93	96	72
Turkmenistan	57	65	52	57	4.18	9.04	99	83	93	54	77	50
Ukraine	61	73	55	64	2.95	7.62	96	–	99	91	98	93
Uzbekistan	63	69	58	61	2.74	9.82	99	–	95	75	78	61
EASTERN MEDITERRANEAN REGION												
Afghanistan	42	42	35	36	0.19	0.22	76	–	63	31	49	29
Bahrain	73	76	64	64	1.09	4.27	98	61	100	–	100	–
Djibouti	53	56	43	43	0.18	0.36	71	–	76	59	88	50
Egypt	66	70	58	60	0.54	2	98	41	99	97	86	58
Iran, Islamic Rep. of	68	73	56	59	0.45	1.31	95	77	99	84	–	–
Iraq	–	–	49	51	0.66	1.25	81	78	97	50	95	48
Jordan	69	73	60	62	2.03	3.24	95	91	99	91	94	87
Kuwait	77	79	67	67	1.53	3.91	99	81	–	–	–	–
Lebanon	68	73	59	62	3.25	1.18	92	87	100	100	100	87
Libyan Arab Jamahiriya	70	75	62	65	1.29	3.6	–	81	–	–	97	96
Morocco	69	73	59	61	0.51	0.78	98	8	99	56	88	52
Oman	71	77	63	65	1.32	3.5	99	71	–	–	97	–
Pakistan	61	62	54	52	0.74	0.46	72	16	96	89	92	41
Qatar	77	78	67	64	2.22	4.94	97	58	100	100	100	100
Saudi Arabia	68	74	60	63	1.37	2.97	96	73	97	–	100	–
Somalia	45	45	36	38	0.04	0.19	35	32	32	27	48	14

	Life expectancy at birth (years)		Healthy life expectancy at birth (years)		Access to health services				Access to improved drinking-water resources (%)		Access to improved sanitation facilities (%)	
					Health care workers		Basic health services					
	Male 2005	Female 2005	Male 2002	Female 2002	Physicians (per 1,000 pop.)	Nurses (per 1,000 pop.)	Immunization rate (DPT3)	ANC coverage at least 4 visits	Urban pop. 2004 (%)	Rural pop. 2004 (%)	Urban pop. 2004 (%)	Rural pop. 2004 (%)
Sudan	57	62	47	50	0.22	0.84	59	75	78	64	50	24
Syrian Arab Republic	70	75	60	63	1.4	1.94	99	51	98	87	99	81
Tunisia	70	75	61	64	1.34	2.87	98	79	99	82	96	65
United Arab Emirates	76	79	64	64	2.02	4.18	94	94	100	100	98	95
Yemen	59	62	48	51	0.33	0.65	86	11	71	65	86	28
WESTERN PACIFIC REGION												
Brunei Darussalam	76	79	65	66	1.01	2.67	99	100	–	–	–	–
Cambodia	51	57	46	49	0.16	0.61	82	9	64	35	53	8
China	71	74	63	65	1.06	1.05	87	–	93	67	69	28
Cook Islands	70	75	61	63	0.78	2.72	99	–	98	88	100	100
Fiji	66	72	57	61	0.34	1.96	75	–	43	51	87	55
Kiribati	62	68	52	56	0.3	2.36	62	88	77	53	59	22
Korea, Rep. of	75	82	65	71	1.57	1.75	96	–	97	71	–	–
Lao People's Democratic Rep.	59	61	47	47	0.59	1.03	49	29	100	100	58	60
Malaysia	69	74	62	65	0.7	1.35	90	–	100	96	95	93
Marshall Islands	60	64	54	56	0.47	2.98	77	–	82	96	93	58
Micronesia, Federated States of	67	70	57	58	0.6	3.83	94	–	95	94	61	14
Mongolia	62	69	53	58	2.63	3.13	99	97	87	30	75	37
Nauru	58	65	53	57	1.45	5.45	80	–	–	–	–	–
Niue	68	74	59	62	1.5	5.5	85	–	100	100	100	100
Palau	68	72	59	60	1.11	1.44	98	–	79	94	96	52
Papua New Guinea	59	63	51	52	0.05	0.53	61	78	88	32	67	41
Philippines	64	71	57	62	0.58	1.69	79	70	87	82	80	59
Samoa	66	70	59	60	0.7	2.02	64	–	90	87	100	100
Singapore	78	82	69	71	1.4	4.24	96	–	100	–	100	–
Solomon Islands	68	72	55	57	0.13	0.8	80	–	94	65	98	18
Tonga	72	70	62	62	0.34	3.16	99	–	100	100	98	96
Tuvalu	61	63	53	53	0.55	2.64	93	–	94	92	93	84
Vanuatu	67	70	58	59	0.11	2.35	66	–	86	52	78	42
Viet Nam	69	74	60	63	0.53	0.56	95	29	99	80	92	50

Indicator definitions

Life expectancy at birth: The average number of years of life that a person can expect to live if they experience the current mortality rate of the population at each age; expressed in years.

Healthy life expectancy at birth: Average number of years that a person can expect to live in "full health" by taking into account years lived in less than full health because of disease or injury.

Physicians/nurses: Ratio of number of physicians/nurses working in the country per 1,000 population.

Immunization rate (DPT3): Percentage of children reaching their first birthday who have been fully immunized against diphtheria, tetanus and whooping cough.

Antenatal clinic (ANC) coverage: Percentage of women who used antenatal care provided by skilled health personnel for reasons related to pregnancy at least four times during pregnancy, as a percentage of live births in a given period.

Access to improved drinking water: Access to an "improved" water supply that provides 20 litres per capita per day at a distance no greater than 1 km.

Access to improved sanitation facilities: Improved sanitation technologies include connection to a public sewer, connection to a septic system, pour/flush latrine, ventilated improved pit latrine, simple pit latrine.

Source

- World Health Organization. 2007. *World Health Statistics 2007*. Geneva.

Table 4 – How many people suffer from poor health during their life?

	Maternal mortality ratio (per 100,000 live births)	Infant mortality rate (per 1,000 live births)	Total population under 5 (thousands)	Probability of dying aged < 5 years per 1,000 live births		% of premature deaths due to communicable diseases
	2000	2005	2005	1990	2005	2002
AFRICAN REGION						
Algeria	140	34	3,160	69	39	50
Angola	1,700	154	2,974	260	260	84
Benin	850	89	1,441	185	150	82
Botswana	100	87	218	58	120	93
Burkina Faso	1,000	96	2,459	210	191	87
Burundi	1,000	114	1,326	190	190	81
Cameroon	730	87	2,453	139	149	81
Cape Verde	150	26	72	60	35	51
Central African Republic	1,100	115	640	168	193	84
Chad	1,100	124	1,867	201	208	85
Comoros	480	53	127	–	71	70
Congo, Democratic Rep. of the	990	129	11,209	205	205	82
Congo, Rep. of the	510	81	750	110	108	79
Côte d'Ivoire	690	118	2,773	157	195	78
Equatorial Guinea	880	123	88	170	205	79
Eritrea	630	50	759	147	78	81
Ethiopia	850	109	13,063	204	164	82
Gabon	420	60	193	92	91	72
Gambia	540	97	231	151	137	75
Ghana	540	68	3,102	122	112	74
Guinea	740	98	1,590	240	150	80
Guinea Bissau	1,100	124	310	253	200	86
Kenya	1,000	79	5,736	97	120	81
Lesotho	550	102	231	101	132	90
Liberia	760	157	631	235	235	83
Madagascar	550	74	3,106	168	119	79
Malawi	1,800	79	2,340	221	125	89
Mali	1,200	120	2,602	250	218	86
Mauritania	1,000	78	526	133	125	79
Mauritius	24	13	98	23	15	11
Mozambique	1,000	100	3,291	235	145	91
Namibia	300	46	268	86	62	83
Niger	1,600	150	2,851	320	256	87
Nigeria	800	100	22,257	230	194	83
Rwanda	1,400	118	1,500	173	203	85
Sao Tome and Principe	–	75	23	118	118	67
Senegal	690	77	1,845	148	136	76
Seychelles	–	12	14	19	13	16
Sierra Leone	2,000	165	958	302	282	86
South Africa	230	55	5,223	60	68	77
Swaziland	370	110	136	110	160	91
Tanzania, United Rep. of	1,500	76	6,045	161	122	85
Togo	570	78	1,014	152	139	79
Uganda	880	79	5,970	160	136	84
Zambia	750	102	2,011	180	182	92
Zimbabwe	1,100	81	1,752	80	132	90
REGION OF THE AMERICAS						
Antigua and Barbuda	–	11	8	–	12	21
Argentina	70	15	3,340	29	18	18

	Maternal mortality ratio (per 100,000 live births)	Infant mortality rate (per 1,000 live births)	Total population under 5 (thousands)	Probability of dying aged < 5 years per 1,000 live births		% of premature deaths due to communicable diseases
	2000	2005	2005	1990	2005	2002
Bahamas	60	13	30	29	15	35
Barbados	95	11	16	17	12	26
Belize	140	15	34	42	17	40
Bolivia	420	52	1,239	125	65	55
Brazil	260	31	18,024	60	33	30
Chile	30	8	1,237	21	10	17
Colombia	130	17	4,726	35	21	25
Costa Rica	25	11	393	18	12	22
Cuba	33	6	682	13	7	10
Dominica	–	13	7	17	15	19
Dominican Republic	150	26	1,003	65	31	56
Ecuador	130	22	1,445	57	25	37
El Salvador	150	23	805	60	27	41
Grenada	–	17	10	37	21	23
Guatemala	240	32	2,020	82	43	60
Guyana	170	47	75	88	63	56
Haiti	680	84	1,147	150	120	84
Honduras	110	31	979	59	40	52
Jamaica	87	17	258	20	20	30
Mexico	83	22	10,857	46	27	27
Nicaragua	230	30	731	68	37	46
Panama	160	19	343	34	24	38
Paraguay	170	20	825	41	23	45
Peru	410	23	2,997	78	27	43
Saint Kitts and Nevis	–	18	4	36	20	26
Saint Lucia	–	12	14	21	14	20
Saint Vincent and the Grenadines	–	17	12	25	20	27
Suriname	110	30	45	48	39	37
Trinidad and Tobago	110	17	90	33	19	40
Uruguay	20	14	282	23	15	12
Venezuela, Bolivarian Rep. of	78	18	2,860	33	21	24
SOUTH EAST ASIA REGION						
Bangladesh	380	54	17,399	149	73	60
Bhutan	420	65	293	166	75	65
India	540	56	120,011	123	74	58
Indonesia	230	28	21,571	91	36	41
Korea, Democratic People's Rep. of	67	42	1,723	55	55	44
Maldives	110	33	46	111	42	55
Myanmar	360	75	4,657	130	105	60
Nepal	740	56	3,639	145	74	64
Sri Lanka	92	12	1,628	32	14	19
Thailand	44	18	5,012	37	21	43
Timor-Leste	660	52	179	177	61	63
EUROPEAN REGION						
Albania	55	16	253	45	18	17
Armenia	55	26	162	54	29	13
Azerbaijan	94	74	602	105	89	36
Belarus	36	10	449	19	12	7
Bosnia and Herzegovina	31	13	186	22	15	7

	Maternal mortality ratio (per 100,000 live births)	Infant mortality rate (per 1,000 live births)	Total population under 5 (thousands)	Probability of dying aged < 5 years per 1,000 live births		% of premature deaths due to communicable diseases
	2000	2005	2005	1990	2005	2002
Bulgaria	32	12	335	18	15	5
Croatia	10	6	207	12	7	5
Cyprus	47	4	49	12	5	12
Czech Republic	9	3	453	13	4	3
Estonia	38	6	64	16	7	6
Georgia	32	41	242	47	45	13
Hungary	11	7	477	17	8	3
Israel	13	5	666	12	6	9
Kazakhstan	210	63	1,075	63	73	16
Kyrgyzstan	110	58	541	80	67	35
Latvia	61	9	101	18	11	7
Lithuania	19	7	150	13	9	4
Macedonia, The Former Yugoslav Rep. of	13	15	117	38	17	8
Malta	–	5	20	11	6	8
Moldova, Rep. of	36	14	207	35	16	11
Poland	10	6	1,811	18	7	4
Romania	58	16	1,054	31	19	11
Russian Federation	65	14	7,225	27	18	8
Serbia and Montenegro	9	12	–	28	15	7
Slovakia	10	7	255	14	8	4
Slovenia	17	3	86	10	4	4
Tajikistan	100	59	834	115	71	49
Turkey	70	26	7,212	82	29	31
Turkmenistan	31	81	488	97	104	35
Ukraine	38	13	1,924	26	17	9
Uzbekistan	24	57	2,841	79	68	30
EASTERN MEDITERRANEAN REGION						
Afghanistan	1,900	165	5,535	260	257	76
Bahrain	33	9	65	19	11	10
Djibouti	730	88	120	175	133	76
Egypt	84	28	8,933	104	33	32
Iran, Islamic Rep. of	76	31	6,035	72	36	22
Iraq	250	102	4,322	50	125	57
Jordan	41	22	732	40	26	31
Kuwait	12	9	241	16	11	18
Lebanon	150	27	322	37	30	18
Libyan Arab Jamahiriya	97	18	636	–	19	31
Morocco	220	36	3,378	89	40	44
Oman	87	10	301	32	12	24
Pakistan	500	79	21,115	130	99	70
Qatar	7	18	67	26	21	16
Saudi Arabia	23	21	3,200	44	26	22
Somalia	1,100	133	1,482	225	225	76
Sudan	590	62	5,216	120	90	60
Syrian Arab Republic	160	14	2,526	39	15	30
Tunisia	120	20	806	52	24	18
United Arab Emirates	54	8	337	15	9	12
Yemen	570	76	3,668	139	102	61

	Maternal mortality ratio (per 100,000 live births)	Infant mortality rate (per 1,000 live births)	Total population under 5 (thousands)	Probability of dying aged < 5 years per 1,000 live births		% of premature deaths due to communicable diseases
	2000	2005	2005	1990	2005	2002
WESTERN PACIFIC REGION						
Brunei Darussalam	37	8	40	11	9	16
Cambodia	450	98	1,835	115	143	72
China	56	23	84,483	49	27	23
Cook Islands	–	17	2	32	20	29
Fiji	75	16	92	22	18	27
Kiribati	–	48	12	88	65	45
Korea, Rep. of	20	5	2,412	9	5	7
Lao People's Democratic Rep.	650	62	895	163	79	71
Malaysia	41	10	2,734	22	12	26
Marshall Islands	–	51	7	92	58	31
Micronesia, Federated States of	–	34	16	58	42	40
Mongolia	110	39	270	108	49	37
Nauru	–	25	2	–	30	19
Niue	–	–	0	–	–	33
Palau	–	10	2	21	11	28
Papua New Guinea	300	55	815	94	74	64
Philippines	200	25	9,863	62	33	45
Samoa	–	24	26	50	29	31
Singapore	15	3	216	9	3	9
Solomon Islands	130	24	72	38	29	49
Tonga	–	20	12	32	24	29
Tuvalu	–	31	1	54	38	34
Vanuatu	–	31	30	62	38	39
Viet Nam	130	16	7,969	53	19	40

Indicator definitions

Maternal mortality ratio: Annual number of deaths of women from pregnancy-related causes per 100,000 live births.

Infant mortality rate: Number of deaths of children under 12 months per 1,000 live births.

Probability of dying aged < 5 years per 1,000 live births: Refers to child mortality risk, which is defined as the probability of dying before age 5; expressed as deaths per 1,000 live births. Also known as under 5 mortality rate.

Years of life lost to communicable diseases: Years of life lost (YLL) are calculated from the number of deaths from communicable diseases multiplied by a standard life expectancy at the age at which death occurs. YLL take into account the age at which deaths occur by giving greater weight to deaths occurring at earlier ages and lower weight than to deaths occurring at later ages. The indicator measures the YLL from a particular cause of death as a percentage of total YLL as a result of premature mortality in the population.

Source

- World Health Organization. 2007. *World Health Statistics 2007.* Geneva.

	Number of refugees		Acute malnutrition of refugees (%)	Demographic change among refugees (%) (1999–2006)	Number of IDPs (thousands)	
	1999	2006			1994–1999*	2000–2006*
Cyprus	120	924	–	670	265	–
Czech Republic	1,200	1,887		–	–	–
Estonia	–	5	–	–	–	–
Georgia	5,200	1,373	–	–74	280	272
Hungary	5,000	8,075	–	62	–	–
Israel	–	–	–	–	–	–
Kazakhstan	14,800	4,412	–	–70	–	–
Kyrgyzstan	10,800	366	–	–97	6	–
Latvia	–	–	–	–	–	–
Lithuania	–	–	–	–	–	–
Macedonia, The Former Yugoslav Rep. of	21,200	1,240	–	–94	–	–
Malta	270	2,404	–	790	–	–
Moldova, Rep. of	–	–	–	–	8	8
Poland	940	6,790	–	622	–	–
Romania	1,200	1,658	–	38	–	–
Russian Federation	80,100	1,425	–	–98	498	561
Serbia and Montenegro[a]	–	98,997	–	–	–	249
Slovakia	–	–	–	–	–	–
Slovenia	4,400	254	–	–94	–	–
Tajikistan	4,500	929	–	–79	17	–
Turkey	2,800	2,633	–	–6	–	–
Turkmenistan	18,500	750	–	–96	–	–
Ukraine	2,700	2,275	–	–16	–	–
Uzbekistan	1,000	1,415	–	42	–	–
EASTERN MEDITERRANEAN REGION						
Afghanistan	–	35	4.0	–	332	759
Bahrain	–	1	–	–	–	–
Djibouti	23,600	9,259	–	–61	–	–
Egypt	6,300	88,022	–	1,297	–	–
Iran, Islamic Rep. of	1,835,700	968,370	–	–47	–	–
Iraq	128,900	44,406	–	–66	–	1,800
Jordan	1,000	500,229	–	49,923	–	–
Kuwait	4,300	50	–	–99	–	–
Lebanon	4,200	20,164	–	380	–	–
Libyan Arab Jamahiriya	–	2,760	–	–	–	–
Morocco	–	–	–	–	–	–
Oman	–	7	–	–	–	–
Pakistan	1,202,000	1,044,462	5.1	–13	–	–
Qatar	–	–	–	–	–	–
Saudi Arabia	5,600	240,772	–	4,200	–	–
Somalia	–	669	–	–	2	400
Sudan	–	196,200	16.2	–	–	1,330
Syrian Arab Republic	6,500	702,209	–	10,703	–	–
Tunisia	–	–	–	–	–	–
United Arab Emirates	–	–	–	–	–	–
Yemen	60,500	95,794	–	58	–	–
WESTERN PACIFIC REGION						
Brunei Darussalam	–	–	–	–	–	–
Cambodia	–	–	–	–	124	–
China	293,300	301,027	–	3	–	–

	Number of refugees		Acute malnutrition of refugees (%)	Demographic change among refugees (%) (1999–2006)	Number of IDPs (thousands)	
	1999	2006			1994–1999*	2000–2006*
Cook Islands	–	–	–	–	–	–
Fiji	–	0	–	–	–	–
Kiribati	–	–	–	–	–	–
Korea, Rep. of	–	–	–	–	–	–
Lao People's Democratic Rep.	–	0	–	–	–	–
Malaysia	50,500	37,170	–	–26	–	–
Marshall Islands	–	–	–	–	–	–
Micronesia, Federated States of	–	2	–	–	–	–
Mongolia	–	5	–	–	–	–
Nauru	–	–	–	–	–	–
Niue	–	–	–	–	–	–
Palau	–	–	–	–	–	–
Papua New Guinea	–	10,183	–	–	–	–
Philippines	–	–	–	–	–	–
Samoa	–	–	–	–	–	–
Singapore	–	–	–	–	–	–
Solomon Islands	–	–	–	–	–	–
Tonga	–	–	–	–	–	–
Tuvalu	–	–	–	–	–	–
Vanuatu	–	720	–	–	–	–
Viet Nam	15,000	2,357	–	–84	–	–

Indicator definitions

Number of refugees: Only countries with more than 1,000 refugees are included.

Demographic change among refugees: Percentage change between numbers in 1999 and numbers in 2006.

Internally displaced people (IDPs): People who have been forced to leave their homes, but who have not crossed an internationally recognized national border.

Acute malnutrition among refugees: The prevalence of wasting – weight-for-height below –2 standard deviations from the median of the reference population – and/or oedema.

Sources

- **Number of refugees/IDPs:** United Nations High Commissioner for Refugees. 2006. *Statistical Yearbook 2006; 2000; 1999; 1994.*
- **Acute malnutrition:** United Nations Standing Committee on Nutrition. 2007. Posted at: http://www.unsystem.org/SCN/Publications/RNIS/rniscountry.html

Notes

* A year span followed by an asterisk indicates that the data are from the most recent year in that period for which data are available.

a Serbia and Montenegro data inclusive of IDPs from Montenegro who are displaced from Serbia (Kosovo) as they were not accorded the same rights as Montenegrin citizens.

Table 6a – How many people suffer from hidden hunger and childhood diseases?

	Hidden hunger in mothers and pregnant women		Hidden hunger in children			Childhood diseases in children under 5		
	Estimated prevalence of iron deficiency anaemia in women aged 15–49 (%)	Estimated annual no. of neural tube birth defects	% in children under 5 with iron deficiency anaemia	Estimated % of children under 6 with vitamin A deficiency	Estimated annual no. of children born mentally impaired (iodine deficiency)	% of all child deaths due to acute respiratory infections	% of all child deaths due to diarrhoeal disease	% of all child deaths due to malaria
	2004	2004	2004	2004	2004	2000	2000	2000
AFRICAN REGION								
Algeria	–	–	–	–	–	13.7	11.9	0.5
Angola	59	1,400	72	55	235,000	24.8	19.1	8.3
Benin	65	550	82	70	10,000	21.1	17.1	27.2
Botswana	31	100	37	30	9,000	1.4	1.1	0.0
Burkina Faso	48	1,230	83	46	180,000	23.3	18.8	20.3
Burundi	60	600	82	44	125,000	22.8	18.2	8.4
Cameroon	32	1,100	58	36	65,000	21.5	17.3	22.8
Cape Verde	–	–	–	–	–	13.3	12.2	4.3
Central African Republic	49	300	74	68	16,000	18.7	14.7	18.5
Chad	56	800	76	45	100,000	22.8	18.1	22.3
Comoros	–	–	–	–	–	16.3	13.6	19.4
Congo, Democratic Rep. of the	54	5,250	58	58	–	23.1	18.1	16.9
Congo, Rep. of the	48	300	55	32	59,000	13.6	11.2	25.7
Côte d'Ivoire	–	–	–	–	–	19.6	14.8	20.5
Equatorial Guinea	–	–	–	–	–	17.3	13.6	24.0
Eritrea	53	300	75	30	16,000	18.6	15.6	13.6
Ethiopia	58	6,000	85	30	685,000	22.3	17.3	6.1
Gabon	32	<100	43	41	11,500	10.7	8.8	28.3
Gambia	53	100	75	64	10,000	15.5	12.2	29.4
Ghana	40	1,300	65	60	120,000	14.6	12.2	33.0
Guinea	43	700	73	40	83,000	20.9	16.5	24.5
Guinea Bissau	53	150	83	31	12,500	23.4	18.6	21.0
Kenya	43	2,000	60	70	105,000	19.9	16.5	13.6
Lesotho	43	100	51	54	11,000	4.7	3.9	0.0
Liberia	44	330	69	38	29,000	23.0	17.3	18.9
Madagascar	42	1,400	73	42	43,000	20.7	16.9	20.1
Malawi	27	1,100	80	59	115,000	22.6	18.1	14.1
Mali	47	1,300	77	47	270,000	23.9	18.3	16.9
Mauritania	42	250	74	17	24,000	22.3	16.2	12.2
Mauritius	–	–	–	–	–	3.9	1.2	0.0
Mozambique	54	1,500	80	26	134,000	21.2	16.5	18.9
Namibia	35	100	42	59	12,000	3.0	2.5	0.0
Niger	47	1,300	57	41	130,000	25.1	19.8	14.3
Nigeria	47	9,500	69	25	370,000	20.1	15.7	24.1
Rwanda	43	700	69	39	46,000	23.2	18.5	4.6
Sao Tome and Principe	–	–	–	–	–	21.2	16.0	0.6
Senegal	43	750	71	61	86,000	20.7	17.1	27.6
Seychelles	–	–	–	–	–	10.1	0.0	0.0
Sierra Leone	68	500	86	47	40,000	25.5	19.7	12.4
South Africa	26	1,500	37	33	160,000	0.9	0.8	0.0
Swaziland	32	<100	47	38	4,000	11.8	9.6	0.2
Tanzania, United Rep. of	45	–	65	37	–	21.1	16.8	22.7
Togo	45	350	72	35	25,000	17.1	13.8	25.3
Uganda	30	2,600	64	66	111,000	21.1	17.2	23.1
Zambia	46	900	63	66	115,000	21.8	17.5	19.4
Zimbabwe	44	800	53	28	35,000	14.7	12.1	0.2
REGION OF THE AMERICAS								
Antigua and Barbuda	–	–	–	–	–	1.5	2.4	0.0
Argentina	–	–	–	–	–	3.4	1.3	0.0

	Hidden hunger in mothers and pregnant women		Hidden hunger in children			Childhood diseases in children under 5		
	Estimated prevalence of iron deficiency anaemia in women aged 15–49 (%)	Estimated annual no. of neural tube birth defects	% in children under 5 with iron deficiency anaemia	Estimated % of children under 6 with vitamin A deficiency	Estimated annual no. of children born mentally impaired (iodine deficiency)	% of all child deaths due to acute respiratory infections	% of all child deaths due to diarrhoeal disease	% of all child deaths due to malaria
	2004	2004	2004	2004	2004	2000	2000	2000
Bahamas	–	–	–	–	–	5.3	0.8	0.0
Barbados	–	–	–	–	–	0.0	0.0	0.0
Belize	–	–	–	–	–	6.9	3.5	0.0
Bolivia	30	380	59	23	13,000	17.1	14.3	0.7
Brazil	21	5,250	45	15	50,000	13.2	12.0	0.5
Chile	–	–	–	–	–	6.2	0.5	0.0
Colombia	–	–	–	–	–	10.4	10.3	0.2
Costa Rica	–	–	–	–	–	4.0	3.0	0.0
Cuba	–	–	–	–	–	4.1	1.3	0.0
Dominica	–	–	–	–	–	0.0	0.0	0.0
Dominican Republic	31	400	25	18	23,000	13.0	11.7	0.6
Ecuador	–	–	–	–	–	12.0	11.0	0.5
El Salvador	34	250	28	17	17,000	13.4	12.4	0.5
Grenada	–	–	–	–	–	9.5	1.6	0.0
Guatemala	20	600	34	21	67,000	15.0	13.1	0.4
Guyana	–	–	–	–	–	5.2	21.4	0.7
Haiti	54	400	66	32	29,000	20.2	16.5	0.7
Honduras	31	300	34	15	24,500	13.8	12.2	0.4
Jamaica	–	–	–	–	–	9.3	9.6	0.0
Mexico	–	–	–	–	–	8.5	5.1	0.0
Nicaragua	40	350	47	9	6,500	13.7	12.2	0.4
Panama	–	–	–	–	–	10.8	10.7	0.2
Paraguay	25	250	52	13	22,000	11.9	10.7	0.3
Peru	32	1,250	50	17	60,000	13.6	12.2	0.4
Saint Kitts and Nevis	–	–	–	–	–	0.0	14.4	0.0
Saint Lucia	–	–	–	–	–	1.3	1.3	0.0
Saint Vincent and the Grenadines	–	–	–	–	–	10.5	0.5	0.0
Suriname	–	–	–	–	–	11.5	13.1	2.4
Trinidad and Tobago	–	–	–	–	–	2.0	1.3	0.0
Uruguay	–	210,000	–	–	–	5.4	2.3	0.0
Venezuela, Bolivarian Rep. of	38	1,200	41	5	60,000	5.9	9.9	0.0
SOUTH EAST ASIA REGION								
Bangladesh	36	8,400	55	28	750,000	17.6	20.0	0.7
Bhutan	55	150	81	32	–	18.8	20.9	0.8
India	51	50,000	75	57	6,600,000	18.5	20.3	0.9
Indonesia	26	6,800	48	26	445,000	14.4	18.3	0.5
Korea, Democratic People's Rep. of	–	–	–	–	–	15.2	18.9	0.7
Maldives	–	–	–	–	–	17.5	20.3	0.6
Myanmar	45	2,300	48	35	205,000	19.3	21.1	9.0
Nepal	62	1,600	65	33	200,000	18.5	20.5	0.8
Sri Lanka	–	–	–	–	–	8.5	13.5	0.4
Thailand	27	2,200	22	22	140,000	11.5	16.2	0.3
Timor-Leste	–	–	–	–	–	19.6	21.9	0.4
EUROPEAN REGION								
Albania	–	–	–	–	–	10.6	10.5	0.4
Armenia	12	<100	24	12	3,500	11.8	10.5	0.5
Azerbaijan	35	225	33	23	22,000	18.4	15.3	1.0
Belarus	–	–	–	–	–	9.0	1.5	0.0

	Hidden hunger in mothers and pregnant women		Hidden hunger in children			Childhood diseases in children under 5		
	Estimated prevalence of iron deficiency anaemia in women aged 15–49 (%)	Estimated annual no. of neural tube birth defects	% in children under 5 with iron deficiency anaemia	Estimated % of children under 6 with vitamin A deficiency	Estimated annual no. of children born mentally impaired (iodine deficiency)	% of all child deaths due to acute respiratory infections	% of all child deaths due to diarrhoeal disease	% of all child deaths due to malaria
	2004	2004	2004	2004	2004	2000	2000	2000
Bosnia and Herzegovina	–	–	–	–	–	2.5	0.6	0.0
Bulgaria	–	–	–	–	–	16.1	2.3	0.0
Croatia	–	–	–	–	–	1.3	0.3	0.0
Cyprus	–	–	–	–	–	1.7	3.2	0.0
Czech Republic	–	–	–	–	–	3.6	0.2	0.0
Estonia	–	–	–	–	–	2.1	1.4	0.0
Georgia	31	100	33	11	11,000	12.5	11.5	0.3
Hungary	–	–	–	–	–	3.9	0.1	0.0
Israel	–	–	–	–	–	0.4	0.6	0.0
Kazakhstan	36	400	49	19	54,000	16.9	14.5	0.8
Kyrgyzstan	31	170	42	18	23,500	16.7	14.1	0.9
Latvia	–	–	–	–	–	1.2	0.0	0.0
Lithuania	–	–	–	–	–	5.3	0.3	0.0
Macedonia, The Former Yugoslav Rep. of	–	–	–	–	–	4.3	5.0	0.0
Malta	–	–	–	–	–	0.0	0.0	0.0
Moldova, Rep. of	–	–	–	–	–	15.5	2.0	0.0
Poland	–	–	–	–	–	2.7	0.1	0.0
Romania	–	–	–	–	–	27.1	2.5	0.0
Russian Federation	–	–	–	–	–	6.3	2.5	0.0
Serbia and Montenegro	–	–	–	–	–	9.4	1.4	0.0
Slovakia	–	–	–	–	–	0.0	0.0	0.0
Slovenia	–	–	–	–	–	19.9	16.4	0.8
Tajikistan	42	300	45	18	43,000	9.1	6.0	0.0
Turkey	33	3,000	23	18	335,000	14.0	12.2	0.5
Turkmenistan	46	200	36	18	11,000	18.8	15.6	0.9
Ukraine	–	–	–	–	–	6.3	1.2	0.0
Uzbekistan	63	800	33	40	136,000	16.8	14.8	0.8
EASTERN MEDITERRANEAN REGION								
Afghanistan	61	2,250	65	53	535,000	24.8	18.9	1.0
Bahrain	–	–	–	–	–	1.4	0.7	0.0
Djibouti	–	–	–	–	–	20.4	16.6	0.8
Egypt	28	3,800	31	7	225,000	14.6	12.8	0.4
Iran, Islamic Rep. of	29	2,100	32	23	125,000	6.4	5.5	0.2
Iraq	–	–	–	–	–	17.6	13.2	0.7
Jordan	–	–	–	–	–	11.7	10.7	0.3
Kuwait	–	–	–	–	–	4.4	0.7	0.0
Lebanon	24	140	21	20	7,500	1.1	1.0	0.0
Libyan Arab Jamahiriya	–	–	–	–	–	8.5	8.4	0.0
Morocco	34	1,000	45	29	–	14.0	12.2	0.4
Oman	–	–	–	–	–	7.2	8.1	0.1
Pakistan	59	11,000	56	35	2,100,000	19.3	14.0	0.7
Qatar	–	–	–	–	–	7.7	8.4	0.0
Saudi Arabia	–	–	–	–	–	6.6	6.2	0.2
Somalia	–	–	–	–	–	23.9	18.7	4.5
Sudan	–	–	–	–	–	15.5	12.9	21.2
Syrian Arab Republic	30	1,000	40	8	40,000	9.9	9.6	0.2
Tunisia	–	–	–	–	–	7.6	7.0	0.2
United Arab Emirates	–	–	–	–	–	4.7	6.3	0.0
Yemen	49	1,800	59	40	143,000	19.8	16.1	7.5

	Hidden hunger in mothers and pregnant women		Hidden hunger in children			Childhood diseases in children under 5		
	Estimated prevalence of iron deficiency anaemia in women aged 15–49 (%)	Estimated annual no. of neural tube birth defects	% in children under 5 with iron deficiency anaemia	Estimated % of children under 6 with vitamin A deficiency	Estimated annual no. of children born mentally impaired (iodine deficiency)	% of all child deaths due to acute respiratory infections	% of all child deaths due to diarrhoeal disease	% of all child deaths due to malaria
	2004	2004	2004	2004	2004	2000	2000	2000
WESTERN PACIFIC REGION								
Brunei Darussalam	–	–	–	–	–	0.7	1.1	0.0
Cambodia	58	950	63	42	85,000	20.6	16.6	0.9
China	21	38,000	8	12	940,000	13.4	11.8	0.4
Cook Islands	–	–	–	–	–	1.1	0.7	0.0
Fiji	–	–	–	–	–	9.2	10.6	0.0
Kiribati	–	–	–	–	–	11.5	21.9	0.7
Korea, Rep. of	–	–	–	–	–	1.8	0.4	0.0
Lao People's Democratic Rep.	48	400	54	42	27,000	19.1	15.6	0.7
Malaysia	–	–	–	–	–	4.0	5.4	0.1
Marshall Islands	–	–	–	–	–	13.5	14.1	0.0
Micronesia, Federated States of	–	–	–	–	–	11.3	8.0	0.0
Mongolia	18	<100	37	29	8,500	17.1	14.5	1.0
Nauru	–	–	–	–	–	30.3	37.8	0.0
Niue	–	–	–	–	–	–	–	–
Palau	–	–	–	–	–	12.4	9.7	0.0
Papua New Guinea	43	350	40	37	–	18.5	15.3	0.8
Philippines	35	4,000	29	23	300,000	13.4	12.0	0.4
Samoa	–	–	–	–	–	10.2	9.7	0.1
Singapore	–	–	–	–	–	9.0	0.4	0.0
Solomon Islands	–	–	–	–	–	9.5	8.8	0.1
Tonga	–	–	–	–	–	7.3	10.0	1.3
Tuvalu	–	–	–	–	–	13.5	13.2	0.0
Vanuatu	–	–	–	–	–	13.0	11.5	0.6
Viet Nam	33	3,300	39	12	180,000	11.5	10.4	0.4

Indicator definitions

Iron deficiency anaemia in women: Percentage of women of reproductive age screened for haemoglobin levels with levels below 110 g/l for pregnant women and 120 g/l for non-pregnant women.

Neural tube birth defects: Defects in a baby's brain and spine, usually a result of lack of folic acid.

Iron deficiency anaemia in children: Percentage of children screened for haemoglobin levels with levels below 8 g/l.

Childhood diseases: Proportion of deaths among children under 5 years from ARIs, diarrhoea or malaria.

Vitamin A deficiency: Estimated percentage of children under 6 with sub-clinical vitamin A deficiency.

Sources

- **Hidden hunger indicators:** United Nations Children's Fund. 2004. *Vitamin and Mineral Deficiency. A Global Progress Report.* New York.
- **Childhood diseases:** World Health Organization. 2007. *World Health Statistics 2007.* Geneva.

Note

Data on vitamin and mineral deficiency are drawn from the best available information. Prevalence data are based on a global review of surveys of vitamin and mineral deficiencies.

Table 6b – How many people suffer from infectious diseases?

	HIV prevalence among adults aged 15–49 years (per 100,000 population)	Estimated number of people living with HIV	Prevalence of tuberculosis (per 100,000 population)	Number of TB cases		% change in the number of TB cases	Estimated proportion of TB patients with HIV (%)
	2005	2005	2005	1990	2005	1990–2005	2005
AFRICAN REGION							
Algeria	82	–	55	11,607	21,336	84	–
Angola	3,281	320,000	333	10,271	37,175	262	19
Benin	1,635	87,000	144	2,084	3,270	57	10
Botswana	23,624	270,000	556	2,938	10,058	242	72
Burkina Faso	2,004	150,000	461	1,497	3,484	133	11
Burundi	3,132	150,000	602	4,575	6,585	44	18
Cameroon	4,899	510,000	206	5,892	21,499	265	26
Cape Verde	–	–	327	221	292	32	–
Central African Republic	9,990	250,000	483	2,124	3,210	51	43
Chad	3,111	180,000	495	2,591	6,311	144	18
Comoros	<500	–	89	140	111	–21	–
Congo, Democratic Rep. of the	2,933	1,000,000	541	21,131	97,075	359	17
Congo, Rep. of the	4,731	120,000	449	591	9,853	1,567	26
Côte d'Ivoire	6,442	750,000	659	7,841	19,681	151	32
Equatorial Guinea	2,857	8,900	355	260	–	–	17
Eritrea	2,180	59,000	515	3,699	3,549	–4	13
Ethiopia	–	–	546	88,634	124,262	40	11
Gabon	6,750	60,000	385	917	2,512	174	33
Gambia	2,091	–	352	–	2,031	–	–
Ghana	2,225	320,000	380	6,407	12,124	89	12
Guinea	1,475	85,000	431	1,988	6,863	245	9
Guinea Bissau	3,483	32,000	293	1,163	1,774	53	20
Kenya	6,125	1,300,000	936	11,788	102,680	771	29
Lesotho	22,684	270,000	588	2,525	10,802	328	69
Liberia	–	–	507	–	3,432	–	17
Madagascar	451	49,000	396	6,261	18,993	203	3
Malawi	12,528	940,000	518	12,395	25,491	106	50
Mali	1,572	130,000	578	2,933	4,697	60	10
Mauritania	629	–	590	5,284	2,162	–59	–
Mauritius	437	–	132	119	125	5	–
Mozambique	14,429	1,800,000	597	15,899	33,231	109	54
Namibia	17,676	230,000	577	2,671	14,920	459	61
Niger	998	79,000	294	5,200	7,873	51	6
Nigeria	3,547	2,900,000	536	20,122	62,598	211	20
Rwanda	3,133	190,000	673	6,387	7,220	13	18
Sao Tome and Principe	–	–	258	17	136	700	–
Senegal	837	–	466	4,977	9,765	96	–
Seychelles	–	–	56	41	14	–66	–
Sierra Leone	1,361	48,000	905	632	6,737	966	9
South Africa	16,579	5,500,000	511	80,400	270,178	236	59
Swaziland	34,457	220,000	1,211	–	8,062	–	79
Tanzania, United Rep. of	5,909	1,400,000	496	22,249	61,022	174	30
Togo	2,879	110,000	753	1,324	2,537	92	17
Uganda	6,304	1,000,000	559	14,740	41,040	178	31
Zambia	15,819	1,100,000	618	16,863	49,576	194	56
Zimbabwe	19,210	1,700,000	631	9,132	50,454	452	62
REGION OF THE AMERICAS							
Antigua and Barbuda	–	–	9	1	6	500	–
Argentina	456	–	51	12,309	9,770	–21	–

	HIV prevalence among adults aged 15–49 years (per 100,000 population)	Estimated number of people living with HIV	Prevalence of tuberculosis (per 100,000 population)	Number of TB cases		% change in the number of TB cases	Estimated proportion of TB patients with HIV (%)
	2005	2005	2005	1990	2005	1990–2005	2005
Bahamas	2,807	6,800	49	46	–	–	17
Barbados	1,236	2,700	12	5	–	–	9
Belize	2,110	3,700	55	57	102	79	13
Bolivia	120	–	280	11,166	9,748	−13	–
Brazil	454	620,000	76	74,570	80,209	8	14
Chile	229	–	16	6,151	2,134	–65	–
Colombia	509	–	66	12,447	10,360	−17	–
Costa Rica	235	–	17	230	534	132	–
Cuba	52	–	11	546	770	41	–
Dominica	–	–	24	6	–	–	–
Dominican Republic	1,036	66,000	116	2,597	5,003	93	6
Ecuador	246	–	202	8,243	4,416	–46	–
El Salvador	770	–	68	2,367	1,794	−24	–
Grenada	–	–	8	0	–	–	–
Guatemala	825	61,000	110	3,813	3,365	−12	6
Guyana	2,072	12,000	194	168	639	280	13
Haiti	3,377	190,000	405	–	14,311	–	19
Honduras	1,392	63,000	99	3,647	3,333	–9	9
Jamaica	1,371	25,000	10	123	90	−27	9
Mexico	244	–	27	14,437	18,524	28	–
Nicaragua	215	–	74	2,944	1,907	−35	–
Panama	755	17,000	46	846	1,637	93	5
Paraguay	338	–	100	2,167	2,075	–4	–
Peru	480	–	206	37,905	33,421	−12	–
Saint Kitts and Nevis	–	–	17	0	–	–	–
Saint Lucia	–	–	22	13	14	8	–
Saint Vincent and the Grenadines	–	–	42	2	7	250	–
Suriname	1,623	5,200	99	82	117	43	11
Trinidad and Tobago	2,538	27,000	13	120	166	38	16
Uruguay	362	–	33	886	622	−30	–
Venezuela, Bolivarian Rep. of	598	–	52	5,457	6,847	25	–
SOUTH EAST ASIA REGION							
Bangladesh	<100	–	406	48,673	123,118	153	–
Bhutan	<100	–	174	1,154	1,007	−13	–
India	747	5,700,000	299	1,519,182	1,156,248	−24	5
Indonesia	106	170,000	262	74,470	254,601	242	1
Korea, Democratic People's Rep. of	–	–	179	–	42,722	–	–
Maldives	–	–	53	152	122	−20	–
Myanmar	982	360,000	170	12,416	107,009	762	7
Nepal	447	–	244	10,142	33,448	230	–
Sri Lanka	<100	–	80	6,666	9,249	39	–
Thailand	1,144	580,000	204	46,510	57,895	24	8
Timor-Leste	–	–	713	–	3,767	–	–
EUROPEAN REGION							
Albania	–	–	28	653	506	−23	–
Armenia	121	–	79	590	2,206	274	–
Azerbaijan	87	–	85	2,620	6,034	130	–
Belarus	242	–	70	3,039	5,308	75	–

	HIV prevalence among adults aged 15–49 years (per 100,000 population)	Estimated number of people living with HIV	Prevalence of tuberculosis (per 100,000 population)	Number of TB cases		% change in the number of TB cases	Estimated proportion of TB patients with HIV (%)
	2005	2005	2005	1990	2005	1990–2005	2005
Bosnia and Herzegovina	–	–	57	4,073	2,111	–48	–
Bulgaria	–	–	41	2,256	3,225	43	–
Croatia	–	–	65	2,576	1,050	–59	–
Cyprus	–	–	5	29	34	17	–
Czech Republic	<100	–	11	1,937	973	–50	–
Estonia	887	1,000	46	423	479	13	7
Georgia	154	–	86	1,537	4,501	193	–
Hungary	<100	–	25	3,588	1,808	–50	–
Israel	–	–	6	234	402	72	–
Kazakhstan	105	–	155	10,969	25,739	135	–
Kyrgyzstan	111	–	133	2,306	6,329	174	–
Latvia	508	–	66	906	1,409	56	–
Lithuania	116	–	63	1,471	2,114	44	–
Macedonia, The Former Yugoslav Rep. of	117	–	33	–	598	–	–
Malta	<500	–	4	13	21	62	–
Moldova, Rep. of	815	–	149	1,728	5,141	198	–
Poland	78	–	29	16,136	8,203	–49	–
Romania	–	–	146	16,256	26,104	61	–
Russian Federation	775	940,000	150	50,641	127,930	153	6
Serbia and Montenegro	–	–	42	4,194	3,208	–24	–
Slovakia	<100	–	20	1,448	710	–51	–
Slovenia	<100	–	15	722	269	–63	–
Tajikistan	123	–	297	2,460	5,460	122	–
Turkey	<100	–	44	24,468	19,744	–19	–
Turkmenistan	<100	–	90	2,325	3,191	37	–
Ukraine	1,036	410,000	120	16,465	39,608	141	8
Uzbekistan	174	–	139	9,414	21,513	129	–
EASTERN MEDITERRANEAN REGION							
Afghanistan	<100	–	288	4,332	21,844	404	–
Bahrain	–	–	43	117	280	139	–
Djibouti	3,017	15,000	1,161	2,100	3,109	48	18
Egypt	<100	–	32	2,142	11,446	434	–
Iran, Islamic Rep. of	133	–	30	9,255	9,422	2	–
Iraq	–	–	76	14,735	9,454	–36	–
Jordan	–	–	6	439	367	–16	–
Kuwait	–	–	28	277	517	87	–
Lebanon	114	–	12	–	391	–	–
Libyan Arab Jamahiriya	–	–	18	442	2,098	375	–
Morocco	88	–	73	27,658	26,269	–5	–
Oman	–	–	11	482	261	–46	–
Pakistan	86	–	297	156,759	137,574	–12	–
Qatar	–	–	65	184	325	77	–
Saudi Arabia	–	–	58	2,415	3,539	47	–
Somalia	870	44,000	286	–	12,904	–	5
Sudan	1,454	350,000	400	212	27,562	12,901	9
Syrian Arab Republic	–	–	46	6,018	4,310	–28	–
Tunisia	115	–	28	2,054	2,079	1	–
United Arab Emirates	–	–	24	285	103	–64	–
Yemen	–	–	136	4,650	9,063	95	–

	HIV prevalence among adults aged 15–49 years (per 100,000 population)	Estimated number of people living with HIV	Prevalence of tuberculosis (per 100,000 population)	Number of TB cases		% change in the number of TB cases	Estimated proportion of TB patients with HIV (%)
	2005	2005	2005	1990	2005	1990–2005	2005
WESTERN PACIFIC REGION							
Brunei Darussalam	<100	–	63	143	163	14	–
Cambodia	1,468	130,000	703	6,501	35,535	447	9
China	62	650,000	208	375,481	894,428	138	0
Cook Islands	–	–	26	1	1	0	–
Fiji	<500	–	30	226	132	–42	–
Kiribati	–	–	426	68	332	388	–
Korea, Rep. of	<100	–	135	63,904	38,290	–40	–
Lao People's Democratic Rep.	103	–	306	1,826	3,777	107	–
Malaysia	391	–	131	11,702	15,342	31	–
Marshall Islands	<100	–	269	–	111	–	–
Micronesia, Federated States of	–	–	123	367	98	–73	–
Mongolia	–	–	206	1,659	4,618	178	–
Nauru	–	–	156	7	11	57	–
Niue	–	–	87	0	–	–	–
Palau	–	–	61	–	10	–	–
Papua New Guinea	1,621	–	475	2,497	12,564	403	–
Philippines	<100	–	450	317,008	137,100	–57	–
Samoa	–	–	27	44	24	–45	–
Singapore	158	–	28	1,591	1,356	–15	–
Solomon Islands	–	–	201	382	397	4	–
Tonga	–	–	32	23	18	–22	–
Tuvalu	–	–	495	23	12	–48	–
Vanuatu	–	–	84	140	76	–46	–
Viet Nam	421	260,000	235	50,203	94,994	89	3

Indicator definitions

HIV prevalence among adults aged 15–49: Number of adults aged 15–49 who are HIV-infected per 100,000 population.

TB prevalence: Number of people reported with TB per 100,000 population.

Estimated proportion of TB patients with HIV: Estimated number of TB patients infected by HIV expressed as a percent.

Sources

- **HIV/AIDS and TB prevalence:** World Health Organization. 2007. World Health Statistics 2007. Geneva.
- **Proportion of TB patients with HIV:** World Health Organization. 2007. *TB/HIV. Networking for Policy Change: an Advocacy Training Manual*. Geneva.

Table 7 – How many people are affected by natural disasters?

	Number of natural disasters 2006	Human impact				Economic losses 2000–2007 (US$ million)
		Total number of affected people 2000–2007	Total number of affected people by natural disaster type 2000–2007			
			Floods	Droughts	Others	
AFRICAN REGION						
Algeria	2	348,478	137,368	0	211,110	5,323,517
Angola	3	565,714	503,694	0	62,020	10,000
Benin	–	19,445	0	0	19,445	0
Botswana	1	166,000	143,736	0	22,264	5,000
Burkina Faso	3	83,237	37,730	0	45,507	0
Burundi	8	1,461,514	61,305	650,000	750,209	0
Cameroon	1	7,102	2,000	0	5,102	0
Cape Verde	–	0	0	0	0	0
Central African Republic	–	45,730	36,743	0	8,987	0
Chad	2	198,794	175,763	0	23,031	1,000
Comoros	2	286,582	0	0	286,582	0
Congo, Democratic Rep. of the	–	–	–	–	–	9,000
Congo, Rep. of the	2	64,567	60,000	0	4,567	0
Côte d'Ivoire	2	4,852	0	0	4,852	0
Equatorial Guinea	–	946	0	0	946	0
Eritrea	–	2,307,013	7,013	2,300,000	0	0
Ethiopia	10	918,245	860,064	0	58,181	9,400
Gabon	–	110	0	0	110	0
Gambia	–	14,522	250	0	14,272	0
Ghana	–	149,614	146,225	0	3,389	0
Guinea	3	225,882	221,200	0	4,682	0
Guinea Bissau	1	26,310	1,000	0	25,310	0
Kenya	10	24,072,497	1,066,808	23,002,000	3,689	538
Lesotho	–	3,835	0	0	3,835	0
Liberia	–	20,865	0	0	20,865	0
Madagascar	3	3,213,766	115,987	0	3,097,779	409,181
Malawi	5	1,050,512	1,005,290	0	45,222	7,700
Mali	4	40,696	38,879	0	1,817	0
Mauritania	2	72,204	69,619	0	2,585	0
Mauritius	1	3,603	0	0	3,603	100,000
Mozambique	4	6,022,291	5,901,266	0	121,025	456,300
Namibia	3	76,753	64,300	0	12,453	8,490
Niger	4	177,572	100,631	0	76,941	0
Nigeria	3	386,385	352,521	0	33,864	12,422
Rwanda	2	930,344	30,516	894,545	5,283	0
Sao Tome and Principe	–	–	–	–	–	–
Senegal	–	261,833	236,769	0	25,064	40,979
Seychelles	1	17,091	0	0	17,091	30,000
Sierra Leone	–	15,726	15,000	0	726	0
South Africa	–	276,225	71,028	0	205,197	420,041
Swaziland	1	282,284	272,000	0	10,284	50
Tanzania, United Rep. of	7	3,049,586	37,293	3,000,000	12,293	–
Togo	1	4,851	2,000	0	2,851	0
Uganda	7	755,943	42,906	700,000	13,037	0
Zambia	2	875,916	862,298	0	13,618	0
Zimbabwe	1	319,197	314,000	0	5,197	277,700
REGION OF THE AMERICAS						
Antigua and Barbuda	–	–	–	–	–	–
Argentina	1	553,406	539,947	0	13,459	1,673,210

	Number of natural disasters 2006	Human impact				Economic losses 2000–2007 (US$ million)
		Total number of affected people 2000–2007	Total number of affected people by natural disaster type 2000–2007			
			Floods	Droughts	Others	
Bahamas	–	10,500	0	0	10,500	1,300,000
Barbados	–	4,880	0	0	4,880	5,000
Belize	–	82,570	0	0	82,570	327,540
Bolivia	2	635,131	593,535	0	41,596	347,000
Brazil	2	1,110,194	489,160	0	621,034	909,370
Chile	1	586,520	531,078	0	55,442	333,900
Colombia	3	1,947,963	1,908,410	0	39,553	10,000
Costa Rica	–	106,720	105,031	0	1,689	47,000
Cuba	2	9,109,275	37,775	0	9,071,500	3,200,200
Dominica	–	275	0	0	275	0
Dominican Republic	–	109,769	82,495	0	27,274	342,230
Ecuador	3	728,158	129,770	0	598,388	179,775
El Salvador	2	1,731,484	3,832	0	1,727,652	2,204,200
Grenada	–	61,650	0	0	61,650	889,000
Guatemala	1	593,064	101,726	0	491,338	988,450
Guyana	1	309,774	309,774	0	0	634,100
Haiti	3	646,753	281,076	0	365,677	73,520
Honduras	1	202,814	6,374	0	196,440	248,500
Jamaica	2	392,502	30,000	0	362,502	1,004,987
Mexico	7	4,161,545	80,940	65,000	4,015,605	9,370,600
Nicaragua	–	89,811	19,046	0	70,765	2,050
Panama	2	67,041	65,453	0	1,588	8,800
Paraguay	1	57,138	2,765	40,000	14,373	0
Peru	3	4,591,881	168,015	0	4,423,866	300,050
Saint Kitts and Nevis	–	–	–	–	–	–
Saint Lucia	–	0	0	0	0	500
Saint Vincent and the Grenadines	–	1,534	0	0	1,534	16,000
Suriname	1	25,000	25,000	0	0	0
Trinidad and Tobago	–	1,760	0	0	1,760	1,000
Uruguay	–	16,712	12,700	0	4,012	275,000
Venezuela, Bolivarian Rep. of	1	134,692	132,962	0	1,730	54,000
SOUTH EAST ASIA REGION						
Bangladesh	8	44,324,694	43,730,663	0	594,031	2,700,000
Bhutan	–	1,000	1,000	0	0	0
India	22	197,482,184	189,290,635	0	8,191,549	18,965,412
Indonesia	21	7,551,578	2,736,764	0	4,814,814	9,214,010
Korea, Democratic People's Rep. of	2	1,190,985	553,404	0	637,581	6,029,400
Maldives	–	27,214	0	0	27,214	470,100
Myanmar	2	164,572	63,750	0	100,822	688
Nepal	4	1,308,857	940,939	0	367,918	6,300
Sri Lanka	1	3,728,862	2,334,443	0	1,394,419	1,351,500
Thailand	4	11,255,123	11,063,918	0	191,205	1,054,663
Timor-Leste	1	12,624	3,558	0	9,066	–
EUROPEAN REGION						
Albania	–	595,110	69,884	0	525,226	17,673
Armenia	–	0	0	0	0	0
Azerbaijan	–	34,794	31,500	0	3,294	65,000
Belarus	1	1,820	0	0	1,820	30,300
Bosnia and Herzegovina	–	290,903	290,100	0	803	0

	Number of natural disasters 2006	Human impact				Economic losses 2000–2007 (US$ million)
		Total number of affected people 2000–2007	Total number of affected people by natural disaster type 2000–2007			
			Floods	Droughts	Others	
Bulgaria	4	12,914	12,200	0	714	477,880
Croatia	1	2,250	2,050	0	200	277,750
Cyprus	–	440	0	0	440	10,000
Czech Republic	3	204,318	204,315	0	3	155,273
Estonia	1	100	0	0	100	130,000
Georgia	1	23,246	3,190	0	20,056	352,156
Hungary	2	46,123	45,823	0	300	148,000
Israel	–	549	0	0	549	2,750
Kazakhstan	–	67,920	31,168	0	36,752	9,162
Kyrgyzstan	2	24,486	2,050	0	22,436	4,160
Latvia	1	102	0	0	102	325,000
Lithuania	1	0	0	0	0	30,000
Macedonia, The Former Yugoslav Rep. of	1	110,103	109,900	0	203	17,163
Malta	–	–	–	–	–	–
Moldova, Rep. of	1	2,607,000	7,000	0	2,600,000	40,184
Poland	2	19,700	19,700	0	0	700,000
Romania	8	198,433	194,932	0	3,501	1,944,790
Russian Federation	5	948,174	860,304	0	87,870	2,648,573
Serbia and Montenegro	3	49,441	48,390	0	1,051	0
Slovakia	2	10,654	330	0	10,324	542,300
Slovenia	–	605	0	0	605	95,000
Tajikistan	3	459,578	426,308	0	33,270	148,670
Turkey	4	733,812	122,635	0	611,177	613,000
Turkmenistan	–	0	0	0	0	0
Ukraine	3	421,650	313,940	0	107,710	243,855
Uzbekistan	–	1,500	1,500	0	0	0
EASTERN MEDITERRANEAN REGION						
Afghanistan	13	910,524	66,098	0	844,426	5,050
Bahrain	–	–	–	–	–	–
Djibouti	1	200,419	100,000	100,000	419	0
Egypt	1	1,387	870	0	517	0
Iran, Islamic Rep. of	4	39,213,511	1,423,407	37,000,000	790,104	4,933,866
Iraq	3	67,910	67,910	0	0	1,300
Jordan	–	237	0	0	237	0
Kuwait	–	1	0	0	1	0
Lebanon	–	17,500	17,000	0	500	0
Libyan Arab Jamahiriya	–	–	–	–	–	0
Morocco	3	316,832	28,367	275,000	13,465	1,553,009
Oman	–	20,083	0	0	20,083	51,000
Pakistan	9	17,625,793	9,205,612	2,200,000	6,220,181	5,743,530
Qatar	–	–	–	–	–	–
Saudi Arabia	–	14,118	13,547	0	571	0
Somalia	9	2,190,561	875,820	1,200,000	114,741	20
Sudan	8	2,897,995	849,836	2,000,000	48,159	186,000
Syrian Arab Republic	1	329,375	0	329,000	375	0
Tunisia	–	27,000	27,000	0	0	0
United Arab Emirates	–	–	–	–	–	–
Yemen	2	6,820	6,341	0	479	0

	Number of natural disasters 2006	Human impact				Economic losses 2000–2007 (US$ million)
		Total number of affected people 2000–2007	Total number of affected people by natural disaster type 2000–2007			
			Floods	Droughts	Others	
WESTERN PACIFIC REGION						
Brunei Darussalam	–	–	–	–	–	–
Cambodia	3	6,625,235	6,625,235	0	0	175,100
China	39	633,081,359	378,252,190	0	254,829,169	48,808,327
Cook Islands	–	1,352	0	0	1,352	0
Fiji	2	35,992	824	0	35,168	34,500
Kiribati	–	–	–	–	–	–
Korea, Rep. of	1	562,073	340,927	0	221,146	7,551,395
Lao People's Democratic Rep.	–	1,062,685	1,053,000	0	9,685	1,000
Malaysia	5	390,865	343,156	0	47,709	80,900
Marshall Islands	–	218	0	0	218	0
Micronesia, Federated States of	–	12,062	0	0	12,062	500
Mongolia	–	1,916,659	5,650	0	1,911,009	80,270
Nauru	–	–	–	–	–	–
Niue	–	702	0	0	702	40,000
Palau	–	–	–	–	–	–
Papua New Guinea	6	91,368	25,193	0	66,175	0
Philippines	20	26,340,777	1,375,815	0	24,964,962	1,382,315
Samoa	–	0	0	0	0	1,500
Singapore	–	2,227	0	0	2,227	0
Solomon Islands	–	7,535	0	0	7,535	0
Tonga	–	16,500	0	0	16,500	51,300
Tuvalu	–	–	–	–	–	0
Vanuatu	1	68,312	3,001	0	65,311	0
Viet Nam	12	13,049,713	8,661,628	0	4,388,085	2,070,525

Indicator definitions

Number of natural disasters, 2006: Includes all hydro-meteorological and geological disasters including drought, flood, earthquake, epidemic, extreme temperature, volcano, wave/surge, wild fires and wind storms.

Number of people affected: People with basic survival needs such as food, water, shelter, sanitation and immediate medical assistance.

Economic losses: Estimated effect on local economy in terms of i) direct costs – damage to infrastructure, crops, housing and ii) indirect costs due to loss of revenues, unemployment and market destabilization.

Source

- Centre for Research on Epidemiology of Disasters. 2007. EM_DAT: The OFDA/CRED International Disaster Database.
  Posted at: http://www.em-dat.net

Table 8 – What solutions are available for addressing hunger and poor health?

	Promoting breastfeeding and safe motherhood		Addressing hidden hunger			
	% of infants exclusively breastfed (0–6 months)	% of infants who receive both breast milk and complementary foods (6–9 months)	Vitamin A supplementation coverage rate 5–59 months (%)	Number of households consuming iodized salt (%)	Children under five years of age with diarrhoea who received oral rehydration therapy (%)	Annual economic loss to micronutrient deficiency (% of GDP)
	1996–2005*	1996–2005*	2004	1998–2005*	1997–2006*	2004
AFRICAN REGION						
Algeria	13	38	–	31	–	–
Angola	11	77	77	65	–	2.1
Benin	38	66	94	28	60.9	1.1
Botswana	34	57	62	34	–	0.6
Burkina Faso	19	38	95	55	62.8	2
Burundi	62	46	94	4	–	2.5
Cameroon	24	79	81	12	56.7	0.8
Cape Verde	57	64	–	100	–	–
Central African Republic	17	77	79	14	–	–
Chad	2	77	84	44	37.5	1.2
Comoros	21	34	7	18	–	–
Congo, Democratic Rep. of the	24	79	81	28	–	0.8
Congo, Rep. of the	19	78	94	–	53.5	1.9
Côte d'Ivoire	5	73	60	16	66.1	–
Equatorial Guinea	24	–	–	67	–	–
Eritrea	52	43	50	32	68.4	1.1
Ethiopia	49	54	52	72	37.1	1.7
Gabon	6	62	–	64	71.6	1.1
Gambia	26	37	27	92	–	1.3
Ghana	53	62	95	72	63.3	1.1
Guinea	27	41	95	32	56.7	1.4
Guinea Bissau	37	36	64	98	–	1.5
Kenya	13	84	63	9	50.6	0.8
Lesotho	36	79	71	9	79.6	0.8
Liberia	35	70	95	–	–	1.2
Madagascar	67	78	89	25	58.1	0.8
Malawi	53	78	57	51	70.1	1.4
Mali	25	32	97	26	65.7	2.7
Mauritania	20	78	95	98	47.7	1.3
Mauritius	21	–	–	100	–	–
Mozambique	30	80	26	46	70.5	1.2
Namibia	19	57	–	37	65.8	0.8
Niger	1	56	–	85	52.9	1.7
Nigeria	17	64	85	3	40.2	0.7
Rwanda	90	69	95	10	31.9	1.1
Sao Tome and Principe	56	53	76	26	–	–
Senegal	34	61	95	59	52.5	1.3
Seychelles	–	–	–	–	–	–
Sierra Leone	4	51	95	77	–	1.4
South Africa	7	46	37	38	89.1	0.4
Swaziland	24	60	86	41	–	0.6
Tanzania, United Rep. of	41	91	94	57	70	–
Togo	18	65	95	33	–	1
Uganda	63	75	68	5	53.1	1
Zambia	40	87	50	23	66.9	1.3
Zimbabwe	33	90	20	7	79.7	0.7

	Promoting breastfeeding and safe motherhood		Addressing hidden hunger			
	% of infants exclusively breastfed (0–6 months)	% of infants who receive both breast milk and complementary foods (6–9 months)	Vitamin A supplementation coverage rate 5–59 months (%)	Number of households consuming iodized salt (%)	Children under five years of age with diarrhoea who received oral rehydration therapy (%)	Annual economic loss to micronutrient deficiency (% of GDP)
	1996–2005*	1996–2005*	2004	1998–2005*	1997–2006*	2004
REGION OF THE AMERICAS						
Antigua and Barbuda	–	–	–	–	–	–
Argentina	–	–	–	10	–	–
Bahamas	–	–	–	–	–	–
Barbados	–	–	–	–	–	–
Belize	24	54	–	10	–	–
Bolivia	54	74	42	10	66.4	0.5
Brazil	–	30	–	12	–	–
Chile	63	47	–	0	–	–
Colombia	47	65	–	8	70.1	–
Costa Rica	35	47	–	3	–	–
Cuba	41	42	–	12	–	–
Dominica	–	–	–	–	–	–
Dominican Republic	10	41	–	82	55	0.4
Ecuador	35	70	–	1	–	–
El Salvador	24	76	–	38	–	0.5
Grenada	39	–	–	–	–	–
Guatemala	51	67	18	33	58.6	0.8
Guyana	11	42	–	–	–	–
Haiti	24	73	–	89	54.9	0.8
Honduras	35	61	40	20	66.8	0.7
Jamaica	–	–	–	0	–	–
Mexico	–	–	–	9	–	–
Nicaragua	31	68	98	3	67.7	0.6
Panama	25	38	–	5	–	–
Paraguay	22	60	–	12	–	0.7
Peru	64	81	–	9	70.6	0.5
Saint Kitts and Nevis	56	–	–	0	–	–
Saint Lucia	–	–	–	–	–	–
Saint Vincent and the Grenadines	–	–	–	–	–	–
Suriname	9	25	–	–	–	–
Trinidad and Tobago	2	19	–	99	–	–
Uruguay	–	–	–	–	–	–
Venezuela, Bolivarian Rep. of	7	50	–	10	–	0.5
SOUTH EAST ASIA REGION						
Bangladesh	36	69	83	30	83.4	0.9
Bhutan	–	–	–	5	–	1.6
India	37	44	51	43	–	1
Indonesia	40	75	73	27	60.6	0.5
Korea, Democratic People's Rep. of	65	31	95	60	–	–
Maldives	10	85	–	56	–	–
Myanmar	15	66	96	40	–	0.7
Nepal	68	66	97	37	46.5	1.5
Sri Lanka	53	–	57	6	–	–
Thailand	4	71	–	37	–	0.4
Timor-Leste	31	82	43	28	–	–
EUROPEAN REGION						
Albania	6	24	–	38	–	–
Armenia	33	57	–	3	59.7	0.3

	Promoting breastfeeding and safe motherhood		Addressing hidden hunger			
	% of infants exclusively breastfed (0–6 months)	% of infants who receive both breast milk and complementary foods (6–9 months)	Vitamin A supplementation coverage rate 5–59 months (%)	Number of households consuming iodized salt (%)	Children under five years of age with diarrhoea who received oral rehydration therapy (%)	Annual economic loss to micronutrient deficiency (% of GDP)
	1996–2005*	1996–2005*	2004	1998–2005*	1997–2006*	2004
Azerbaijan	7	39	14	74	–	0.7
Belarus	–	–	–	45	–	–
Bosnia and Herzegovina	6	–	–	38	–	–
Bulgaria	–	–	–	2	–	–
Croatia	23	–	–	10	–	–
Cyprus	–	–	–	–	–	–
Czech Republic	–	–	–	–	–	–
Estonia	–	–	–	–	–	–
Georgia	18	12	–	32	–	0.5
Hungary	–	–	–	–	–	–
Israel	–	–	–	–	–	–
Kazakhstan	36	73	–	17	52.6	0.6
Kyrgyzstan	24	77	95	58	73.5	0.9
Latvia	–	–	–	–	–	–
Lithuania	–	–	–	–	–	–
Macedonia, The Former Yugoslav Rep. of	37	8	–	6	–	–
Malta	–	–	–	–	–	–
Moldova, Rep. of	46	66	–	41	–	–
Poland	–	–	–	–	–	–
Romania	16	41	–	47	–	–
Russian Federation	–	–	–	65	–	–
Serbia and Montenegro	–	–	–	–	–	–
Slovakia	–	–	–	–	–	–
Slovenia	–	–	–	–	–	–
Tajikistan	41	91	98	72	–	1.2
Turkey	21	38	–	36	–	0.7
Turkmenistan	13	71	–	0	–	0.7
Ukraine	22	–	–	68	–	–
Uzbekistan	19	49	86	43	–	1.2
EASTERN MEDITERRANEAN REGION						
Afghanistan	–	–	96	72	–	2.3
Bahrain	34	65	–	–	–	–
Djibouti	–	–	–	–	–	–
Egypt	38	67	–	22	35.7	0.5
Iran, Islamic Rep. of	44	–	–	6	–	0.3
Iraq	12	51	–	60	–	–
Jordan	27	70	–	12	63.9	–
Kuwait	12	26	–	–	–	–
Lebanon	27	35	–	8	–	0.4
Libyan Arab Jamahiriya	–	–	–	10	–	–
Morocco	–	–	–	–	54	0.2
Oman	–	92	95	39	–	–
Pakistan	16	31	95	83	–	1.7
Qatar	12	48	–	–	–	–
Saudi Arabia	31	60	–	–	–	–
Somalia	9	13	6	–	–	–
Sudan	16	47	70	99	–	–
Syrian Arab Republic	81	50	–	21	–	0.5
Tunisia	47	–	–	3	–	–

	Promoting breastfeeding and safe motherhood		Addressing hidden hunger			
	% of infants exclusively breastfed (0–6 months)	% of infants who receive both breast milk and complementary foods (6–9 months)	Vitamin A supplementation coverage rate 5–59 months (%)	Number of households consuming iodized salt (%)	Children under five years of age with diarrhoea who received oral rehydration therapy (%)	Annual economic loss to micronutrient deficiency (% of GDP)
	1996–2005*	1996–2005*	2004	1998–2005*	1997–2006*	2004
United Arab Emirates	34	52	–	–	–	–
Yemen	12	76	20	70	64.6	1.3
WESTERN PACIFIC REGION						
Brunei Darussalam	–	–	–	–	–	–
Cambodia	12	72	72	86	74.1	1.4
China	51	32	–	7	–	0.2
Cook Islands	19	–	–	–	–	–
Fiji	47	–	–	69	–	–
Kiribati	80	–	58	–	–	–
Korea, Rep. of	–	–	–	–	–	–
Lao People's Democratic Rep.	23	10	48	25	–	1.1
Malaysia	29	–	–	–	–	–
Marshall Islands	63	–	24	–	–	–
Micronesia, Federated States of	60	–	74	–	–	–
Mongolia	51	55	93	25	–	0.6
Nauru	–	–	–	–	–	–
Niue	–	–	–	–	–	–
Palau	59	–	–	–	–	–
Papua New Guinea	59	74	32	–	–	0.5
Philippines	34	58	85	44	58.9	0.7
Samoa	–	–	–	–	–	–
Singapore	–	–	–	–	–	–
Solomon Islands	65	–	–	–	–	–
Tonga	62	–	–	–	–	–
Tuvalu	–	–	–	–	–	–
Vanuatu	50	–	–	–	–	–
Viet Nam	15	–	95	17	74	0.6

Indicator definitions

Exclusive breastfeeding: Children fed exclusively on breast milk.

Infants who receive breast milk and complementary foods: Complementary foods are solid or semi-solid foods, given in addition to breast milk.

Children under 5 with diarrhoea who received ORT: Proportion of children aged 0–59 months who had diarrhoea in the preceding two weeks and were treated with ORT or a similar solution.

Economic loss caused by micronutrient deficiencies: Estimated economic losses in terms of death, disability and lost productivity.

Note

A year span followed by an asterisk indicates that the data are from the most recent year in that period for which data are available.

Sources

- **Breastfeeding and safe motherhood:** United Nations Children's Fund. 2006. *The State of the World's Children 2007 – Women and Children. The Double Dividend of Gender Equality.* New York.

  UNICEF End Decade Database – Integrated Management of Childhood Illness (IMCI).
- **Losses to vitamin deficiency:** United Nations Children's Fund. 2004. *Vitamin and Mineral Deficiency. A Global Progress Report.* New York.
- **Vitamin A coverage and consumption of iodized salt:** United Nations Children's Fund. 2006. *The State of the World's Children 2007 – Women and Children. The Double Dividend of Gender Equality*. New York.
- **Children under 5 with diarrhoea who received ORT:** World Health Organization. 2007. *World Health Statistics 2007*. Geneva.
- **Economic loss caused by micronutrient deficiency:** United Nations Children's Fund. 2004. *Vitamin and Mineral Deficiency. A Global Progress Report.* New York.

Table 9 – What resources are allocated to reduce hunger and poor health?

	National health sector expenditures				Trends of Official Development Assistance (ODA)		Food assistance	
	Out-of-pocket expenditure as percentage of private expenditure on health (%)	Per capita total expenditure on health (US$)	Per capita government expenditure on health (US$)	Total expenditure on health as percentage of gross domestic product (%)	Net received ODA (millions)	ODA received as % of GNI	Global food assistance deliveries (thousand tons)	WFP's shares of total food assistance deliveries (%)
	2004	2004	2004	2004	2005	2005	2006	2006
AFRICAN REGION								
Algeria	94.6	93.9	68.1	3.6	371	0.38	20	100
Angola	100	25.5	20.3	1.9	442	1.73	23	79
Benin	99.9	24.2	12.4	4.9	349	8.20	15	42
Botswana	27.9	328.6	206.7	6.4	71	0.80	–	–
Burkina Faso	97.9	24.2	13.3	6.1	660	12.78	43	28
Burundi	100	3	0.8	3.2	365	46.79	74	89
Cameroon	94.5	50.7	14.2	5.2	414	2.50	10	36
Cape Verde	99.8	97.8	74.1	5.2	161	17.05	25	7
Central African Republic	95.4	13.2	4.9	4.1	95	6.97	14	72
Chad	95.8	19.6	7.2	4.2	380	8.55	70	96
Comoros	100	13.2	7.5	2.8	25	6.64	–	–
Congo, Democratic Rep. of the	100	4.7	1.3	4	1,828	27.54	90	74
Congo, Rep. of the	100	27.6	13.6	2.5	1,449	36.82	12	22
Côte d'Ivoire	88.7	33	7.9	3.8	119	0.78	27	100
Equatorial Guinea	75.1	168.2	129.6	1.6	39	–	–	–
Eritrea	100	9.9	3.9	4.5	355	36.32	42	–
Ethiopia	78.3	5.6	2.9	5.3	1,937	17.39	619	68
Gabon	100	231.3	159.1	4.5	54	0.74	–	–
Gambia	68.2	18.5	5	6.8	58	13.06	14	35
Ghana	78.2	27.2	11.5	6.7	1,120	10.63	59	13
Guinea	99.5	21.8	2.9	5.3	182	6.89	25	56
Guinea Bissau	90	8.7	2.4	4.8	79	27.33	9	73
Kenya	81.9	20.1	8.6	4.1	768	4.27	385	93
Lesotho	18.2	49.4	41.6	6.5	69	3.84	16	95
Liberia	98.5	8.6	5.5	5.6	236	54.12	49	86
Madagascar	52.5	7.3	4.3	3	929	18.75	44	15
Malawi	35.2	19.3	14.4	12.9	575	28.37	201	39
Mali	99.5	23.8	11.7	6.6	691	14.08	47	42
Mauritania	100	14.5	10.1	2.9	190	10.43	29	57
Mauritius	80.8	222.3	121.6	4.3	32	0.50	–	–
Mozambique	38.5	12.3	8.4	4	1,286	20.78	129	31
Namibia	18.1	189.8	131	6.8	123	1.99	7	59
Niger	85.1	8.6	4.5	4.2	515	15.17	97	63
Nigeria	90.4	23	7	4.6	6,437	7.41	–	–
Rwanda	36.9	15.5	8.8	7.5	576	27.39	44	75
Sao Tome and Principe	100	47.8	41.2	11.5	32	58.56	1	100
Senegal	94.5	39.4	15.9	5.9	689	8.44	31	41
Seychelles	62.5	534.4	402.5	6.1	19	2.83	–	–
Sierra Leone	100	6.6	3.9	3.3	343	29.58	29	51
South Africa	17.2	390.2	157.5	8.6	700	0.30	–	–
Swaziland	40.2	145.8	93.1	6.3	46	1.67	13	100
Tanzania, United Rep. of	83.2	12	5.2	4	1,505	12.48	232	90
Togo	84.9	17.9	3.7	5.5	87	4.00	86	100
Uganda	51.3	19	6.2	7.6	1,198	14.02	1	100
Zambia	71.4	29.6	16.2	6.3	945	14.21	89	100
Zimbabwe	48.7	27.2	12.5	7.5	368	11.55	176	100
REGION OF THE AMERICAS								
Antigua and Barbuda	100	485.3	342.7	4.8	7	–	–	–
Argentina	48.7	382.9	173.5	9.6	100	0.06	–	–
Bahamas	40.3	1,211	606.7	6.8	–	–	–	–

	National health sector expenditures				Trends of Official Development Assistance (ODA)		Food assistance	
	Out-of-pocket expenditure as percentage of private expenditure on health (%)	Per capita total expenditure on health (US$)	Per capita government expenditure on health (US$)	Total expenditure on health as percentage of gross domestic product (%)	Net received ODA (millions)	ODA received as % of GNI	Global food assistance deliveries (thousand tons)	WFP's shares of total food assistance deliveries (%)
	2004	2004	2004	2004	2005	2005	2006	2006
Barbados	78.6	744.8	473.3	7.1	– 2	–	0.15	100
Belize	100	200.5	107.8	5.1	13	–	–	–
Bolivia	82.5	65.8	40	6.8	583	6.50	109	9
Brazil	64.2	289.5	156.6	8.8	192	0.03	–	–
Chile	45.9	359	168.7	6.1	152	0.14	–	–
Colombia	49	168.3	144.7	7.8	511	0.44	19	90
Costa Rica	88.7	289.7	223.2	6.6	30	–	–	–
Cuba	74.5	229.8	201.8	6.3	88	–	5	100
Dominica	100	215.1	153.3	5.9	15	–	–	–
Dominican Republic	73.1	148.1	46.8	6	77	–	–	–
Ecuador	85.4	127.3	51.9	5.5	210	0.61	31	4
El Salvador	94.2	183.8	81.6	7.9	199	–	19	24
Grenada	100	292.9	213.2	6.9	45	–	–	–
Guatemala	90.5	126.9	52	5.7	254	–	79	41
Guyana	100	55.9	46.7	5.3	137	18.64	–	–
Haiti	69.6	33	12.7	7.6	515	–	116	21
Honduras	84.3	77.1	42.3	7.2	681	–	72	3
Jamaica	63.6	175.6	95.3	5.2	36	–	–	–
Mexico	94.4	424.3	196.8	6.5	189	–	0	–
Nicaragua	95.9	67.1	31.6	8.2	740	–	42	28
Panama	82.5	342.7	229.3	7.7	20	–	–	–
Paraguay	72.2	88.4	29.8	7.7	51	0.63	–	–
Peru	79.2	103.7	48.6	4.1	398	0.54	62	4
Saint Kitts and Nevis	100	500.4	316	5.2	4	–	–	–
Saint Lucia	100	232.4	151.1	5	11	–	–	–
Saint Vincent and the Grenadines	100	210.2	132.8	6.1	5	–	–	–
Suriname	60.2	194.2	89.3	7.8	44	3.78	–	–
Trinidad and Tobago	88.5	329.3	128.1	3.5	– 2	–	–	–
Uruguay	31.1	314.7	136.9	8.2	15	0.09	–	–
Venezuela (Bolivarian Republic of)	88.3	195.6	82.1	4.7	49	0.04	–	–
SOUTH EAST ASIA REGION								
Bangladesh	88.3	13.7	3.8	3.1	1,321	2.10	209	34
Bhutan	100	15.4	9.9	4.6	90	10.96	3	100
India	93.8	31.4	5.4	5	1,724	0.22	143	39
Indonesia	74.7	32.5	11.1	2.8	2,524	0.91	91	88
Korea, Democratic People's Rep. of	100	0.3	0.2	3.5	81	–	223	8
Maldives	100	180.1	146.7	7.7	67	8.54	6	–
Myanmar	99.4	4.5	0.6	2.2	145	–	27	85
Nepal	88.1	14.1	3.7	5.6	428	5.81	66	75
Sri Lanka	84	42.5	19.4	4.3	1,189	5.13	42	100
Thailand	74.7	88.1	57	3.5	–171	–0.10	–	–
Timor-Leste	25.6	43.5	34.3	11.2	185	26.74	11	99
EUROPEAN REGION								
Albania	99.8	157.1	69.3	6.7	319	3.73	–	–
Armenia	89.2	63	16.5	5.4	193	3.91	9	81
Azerbaijan	93.6	37.2	9.3	3.6	223	2.04	90	6
Belarus	72.7	146.7	109.8	6.2	54	0.18	–	–
Bosnia and Herzegovina	100	197.6	97.5	8.3	546	5.71	–	–
Bulgaria	98	250.8	144.4	8	–	–	–	–
Croatia	93.8	609.4	493.8	7.7	125	0.35	–	–

	National health sector expenditures				Trends of Official Development Assistance (ODA)		Food assistance	
	Out-of-pocket expenditure as percentage of private expenditure on health (%)	Per capita total expenditure on health (US$)	Per capita government expenditure on health (US$)	Total expenditure on health as percentage of gross domestic product (%)	Net received ODA (millions)	ODA received as % of GNI	Global food assistance deliveries (thousand tons)	WFP's shares of total food assistance deliveries (%)
	2004	2004	2004	2004	2005	2005	2006	2006
Cyprus	93.4	1,109.4	491.2	5.8	–	–	–	–
Czech Republic	95.5	770.8	687.2	7.3	–	–	–	–
Estonia	88.8	462.8	351.7	5.3	–	–	–	–
Georgia	87.2	60	16.4	5.3	310	4.70	18	48
Hungary	88	800.2	573	7.9	–	–	–	–
Israel	75	1,533.5	1,073.4	8.7	–	–	–	–
Kazakhstan	100	109.1	65.3	3.8	229	0.45	–	–
Kyrgyzstan	94.3	23.7	9.7	5.6	268	11.37	0	–
Latvia	98.3	417.8	236.7	7.1	–	–	–	–
Lithuania	96.8	423.8	318.1	6.5	–	–	–	–
Macedonia, The Former Yugoslav Rep. of	100	211.6	150.3	8	230	4.04	–	–
Malta	90.2	1,239.4	942.8	9.2	–	–	–	–
Moldova, Rep. of	96	45.7	26	7.4	192	5.85	28	–
Poland	89.6	410.7	281.9	6.2	–	–	–	–
Romania	93.4	177.6	117.4	5.1	–	–	–	–
Russian Federation	76.7	244.7	150.1	6	–	–	17	87
Serbia and Montenegro	88.2	–	–	10.1	1,132	4.26	–	–
Slovakia	73.1	565.1	416.9	7.2	–	–	–	–
Slovenia	39.5	1,438.2	1,087	8.7	–	–	–	–
Tajikistan	97.3	14.1	3	4.4	241	10.79	68	38
Turkey	69.1	324.8	234.7	7.7	464	0.13	–	–
Turkmenistan	100	124.4	85.7	4.8	28	–	–	–
Ukraine	90.5	89.7	50.9	6.5	410	0.51	–	–
Uzbekistan	96.2	23.2	10.8	5.1	172	1.27	–	–
EASTERN MEDITERRANEAN REGION								
Afghanistan	97.7	13.5	2.3	4.4	2,775	38.55	150	94
Bahrain	69.3	619.8	416.6	4	–	–	–	–
Djibouti	98.6	53.1	36.7	6.3	79	10.09	11	82
Egypt	94.3	66	25.2	6.1	926	–	29	16
Iran, Islamic Rep. of	94.8	157.8	75.4	6.6	104	0.05	1	44
Iraq	100	57.7	45.3	5.3	21,654	–	4	67
Jordan	73.8	199.9	96.7	9.8	622	4.70	109	2
Kuwait	90.4	632.8	490.8	2.8	–	–	–	–
Lebanon	82.2	670.2	183.6	11.6	243	1.16	35	38
Libyan Arab Jamahiriya	100	195.4	146.3	3.8	–	–	–	–
Morocco	76	82.2	28.2	5.1	652	–	–	–
Oman	57.1	294.6	239.9	3	31	–	–	–
Pakistan	98	13.6	2.7	2.2	1,666	1.54	92	86
Qatar	86.4	992.4	759	2.4	–	–	–	–
Saudi Arabia	26.4	412	315.1	3.9	26	0.01	–	–
Somalia	–	–	–	–	236	–	179	47
Sudan	98.1	24.7	8.7	4.1	1,829	7.10	497	89
Syrian Arab Republic	100	57.8	27.4	4.7	78	0.30	3	100
Tunisia	83	175	91.2	6.2	376	–	–	–
United Arab Emirates	71	711.2	497.4	2.9	–	–	–	–
Yemen	95.5	33.9	13	5	336	2.61	114	17
WESTERN PACIFIC REGION								
Brunei Darussalam	100	473.2	377	3.2	–	–	–	–
Cambodia	85.4	23.6	6.1	6.7	538	10.39	25	91
China	86.5	70.1	26.6	4.7	1,757	0.08	–	–

	National health sector expenditures				Trends of Official Development Assistance (ODA)		Food assistance	
	Out-of-pocket expenditure as percentage of private expenditure on health (%)	Per capita total expenditure on health (US$)	Per capita government expenditure on health (US$)	Total expenditure on health as percentage of gross domestic product (%)	Net received ODA (millions)	ODA received as % of GNI	Global food assistance deliveries (thousand tons)	WFP's shares of total food assistance deliveries (%)
	2004	2004	2004	2004	2005	2005	2006	2006
Cook Islands	100	334.9	292.7	3.5	8	–	–	–
Fiji	100	147.6	92	4.6	64	2.28	–	–
Kiribati	100	111.9	104.1	13.7	28	19.95	–	–
Korea, Rep. of	80.4	776.9	408.5	5.5	–	–	–	–
Lao People's Democratic Rep.	90.3	16.8	3.4	3.9	296	11.17	12	46
Malaysia	74.1	180.1	105.9	3.8	32	0.03	–	–
Marshall Islands	100	271.9	263.8	15.2	57	31.43	–	–
Micronesia, Federated States of	40	155.8	133.6	7.6	106	43.89	–	–
Mongolia	92.3	37.3	24.8	6	212	11.57	34	–
Nauru	100	346.8	253.2	8.1	9	–	–	–
Niue	100	1,225.2	1,210.8	15.1	21	–	–	–
Palau	100	655.8	598.1	9.7	23	15.76	–	–
Papua New Guinea	46.4	30.4	25.7	3.6	266	6.64	–	–
Philippines	77.9	36.1	14.4	3.4	562	0.53	33	27
Samoa	78	108.9	83.6	5.3	44	11.23	–	–
Singapore	96.9	942.9	321	3.7	–	–	–	–
Solomon Islands	55.9	34.8	32.9	5.9	198.24	70.51	–	–
Tonga	84.9	117.1	93.1	6.3	32	13.01	–	–
Tuvalu	16.5	355.3	336.7	16.6	9	–	–	–
Vanuatu	57.5	58	44.5	4.1	39	11.98	–	–
Viet Nam	88	30	8.1	5.5	1,905	3.69	–	–

Indicator definitions

Out-of-pocket expenditure as percentage of private expenditure on health: Ratio between household out-of-pocket spending and total private health expenditure, which is defined as the sum of expenditures on health by pre-paid plans and risk-pooling arrangements, firms' expenditure on health, non-profit institutions serving mainly households and household out-of-pocket spending.

Government expenditure on health: The sum of outlays by government entities to purchase healthcare services and goods. It comprises outlays on health by all levels of government and social-security agencies, and direct expenditure by para-statal organizations and public companies. Expenditures on health include final consumption, subsidies to producers and transfers to households mainly as reimbursements for medical and pharmaceutical bills. It includes recurrent and investment expenditures, including capital transfers, made during the year and external resources mainly as grants passing through the government or loans channelled through the national budget. Expressed in US dollars.

Total expenditure on health: This is the sum of government health expenditure and private health expenditure in a given year, expressed in US dollars. It comprises the outlays earmarked for health maintenance, restoration or enhancement of the health status of the population, paid for in cash or in kind.

Gross domestic product (GDP): This is the value of all goods and services provided in a country by residents and non-residents. It corresponds to the total sum of expenditure – consumption and investment – of private and government agents of the economy during the reference year.

Net received official development assistance (ODA): Total net official development assistance flows from DAC countries, multilateral organizations and non-DAC countries.

Gross national income (GNI): GNI measured at market prices as the aggregate value of the balances of gross primary incomes for all sectors; GNI is identical to gross national product (GNP) as previously used in national accounts.

Global food assistance deliveries: 2006 international food assistance deliveries based on shipments during the year expressed in thousands of metric tons; cereals are expressed in grain equivalent. Does not include national food assistance, which is assistance funded from the recipient country.

WFP's share of total food assistance deliveries: Tonnage of total food assistance deliveries by WFP in 2006.

Sources

- **ODA receipts, GNI:** Organisation for Economic Co-operation and Development. 2006. *Statistical Annex of the 2006 Development Co-operation Report*. Paris. Posted at: www.oecd.org/dac/stats/dac/dcrannex
- **Health sector expenditures:** World Health Organization. 2007. *World Health Statistics 2007*. Geneva.
- **Food assistance:** World Food Programme. 2007. International Food Aid Information System (INTERFAIS). Rome. Posted at www.wfp.org/interfais

Table 10 – Progress towards achieving the MDGs by 2015

	MDG 1: Eradicate extreme poverty and hunger		MDG 4: Reduce child mortality		MDG 5: Reduce maternal mortality
	Target 2: Halve, between 1990 and 2015, the proportion of people suffering from hunger		Target 5: Reduce by two thirds, between 1990 and 2015, the under 5 mortality rate		Target 6: Reduce by three quarters, between 1990 and 2015, the maternal mortality ratio
	Children under 5 moderately or severely underweight	Population undernourished	Infant mortality (0–1 year)	Children under 5 mortality	Maternal mortality
	1990–2006**	1990–2003**	1990–2005	1990–2005	1990–2000
AFRICAN REGION					
Algeria	–	0.00	0.56	0.65	0.17
Angola	0.65	0.69	0.00	0.00	-0.18
Benin	–	0.60	0.30	0.28	0.19
Botswana	–	-0.61	-1.40	-1.60	0.80
Burkina Faso	-0.15	0.38	0.23	0.14	-0.10
Burundi	-0.38	-0.79	0.00	0.00	0.31
Cameroon	0.00	0.48	-0.04	-0.11	-0.44
Cape Verde	–	–	0.63	0.63	–
Central African Republic	0.40	0.20	-0.19	-0.22	-0.76
Chad	0.25	0.86	-0.09	-0.04	0.36
Comoros	-0.70	–	0.60	0.61	0.66
Congo, Democratic Rep. of the	0.05	-2.65	0.00	0.00	-0.18
Congo, Rep. of the	1.01	0.74	0.04	0.03	0.57
Côte d'Ivoire	0.47	0.44	-0.22	-0.36	0.20
Equatorial Guinea	–	–	-0.29	-0.31	-0.10
Eritrea	0.32	-0.15	0.65	0.70	0.73
Ethiopia	0.50	0.49	0.25	0.29	0.52
Gabon	–	1.00	0.00	0.02	0.21
Gambia	0.82	-0.45	0.09	0.17	0.68
Ghana	0.62	1.35	0.14	0.12	-31.39
Guinea	–	0.77	0.49	0.56	0.72
Guinea Bissau	–	–	0.28	0.31	-0.28
Kenya	0.54	0.41	-0.35	-0.36	-0.72
Lesotho	0.10	0.59	-0.32	-0.15	0.13
Liberia	–	-0.88	0.00	0.00	-0.48
Madagascar	0.20	-0.17	0.42	0.44	-0.16
Malawi	0.46	0.64	0.69	0.72	-2.95
Mali	–	0.07	0.21	0.19	0.00
Mauritania	0.72	0.67	0.12	0.09	-0.10
Mauritius	2.00	0.00	0.57	0.52	1.07
Mozambique	0.43	0.64	0.55	0.57	0.44
Namibia	–	0.65	0.35	0.42	0.25
Niger	-0.05	0.44	0.32	0.30	-0.44
Nigeria	0.47	0.62	0.25	0.23	0.27
Rwanda	0.62	0.33	-0.22	-0.26	-0.10
Sao Tome and Principe	–	–	0.00	0.00	–
Senegal	0.66	0.00	0.22	0.12	0.57
Seychelles	2.00	–	0.44	0.47	–
Sierra Leone	–	-0.17	0.09	0.10	–
South Africa	–	–	-0.33	-0.20	0.00
Swaziland	–	-0.71	-0.62	-0.68	0.45
Tanzania, United Rep. of	0.84	-0.38	0.38	0.36	-1.26
Togo	–	0.48	0.17	0.13	–
Uganda	0.35	0.42	0.23	0.23	0.36
Zambia	-0.27	0.04	-0.01	-0.02	0.27
Zimbabwe	0.00	0.00	-0.79	-0.98	-1.24
REGION OF THE AMERICAS					
Antigua and Barbuda	–	–	–	–	–
Argentina	-0.42	–	0.63	0.57	0.40

	MDG 1: Eradicate extreme poverty and hunger		MDG 4: Reduce child mortality		MDG 5: Reduce maternal mortality
	Target 2: Halve, between 1990 and 2015, the proportion of people suffering from hunger		Target 5: Reduce by two thirds, between 1990 and 2015, the under 5 mortality rate		Target 6: Reduce by three quarters, between 1990 and 2015, the maternal mortality ratio
	Children under 5 moderately or severely underweight	Population undernourished	Infant mortality (0–1 year)	Children under 5 mortality	Maternal mortality
	1990–2006**	1990–2003**	1990–2005	1990–2005	1990–2000
Bahamas	–	–	0.69	0.72	0.53
Barbados	–	–	0.32	0.38	–1.61
Belize	–	–	0.92	0.98	–
Bolivia	0.96	0.36	0.62	0.72	0.47
Brazil	0.94	0.67	0.57	0.68	–0.24
Chile	1.36	1.00	0.79	0.79	0.72
Colombia	0.99	0.35	0.65	0.63	–0.40
Costa Rica	2.00	0.67	0.47	0.50	0.73
Cuba	–	–	0.68	0.69	0.87
Dominica	–	–	0.20	0.18	–
Dominican Republic	1.18	0.00	0.72	0.78	–0.48
Ecuador	1.25	0.75	0.73	0.84	0.18
El Salvador	1.20	0.17	0.77	0.83	0.67
Grenada	–	–	0.65	0.65	1.33
Guatemala	0.93	–0.88	0.70	0.71	–0.27
Guyana	0.70	1.14	0.40	0.43	–
Haiti	0.96	0.55	0.26	0.30	0.43
Honduras	1.04	0.09	0.44	0.48	0.67
Jamaica	0.65	0.57	0.00	0.00	0.37
Mexico	1.52	0.00	0.61	0.62	0.33
Nicaragua	0.58	0.20	0.63	0.68	–0.58
Panama	1.20	–0.38	0.44	0.44	–2.55
Paraguay	–	0.33	0.59	0.66	–0.08
Peru	1.03	1.43	0.93	0.99	–0.62
Saint Kitts and Nevis	–	–	0.60	0.67	–
Saint Lucia	–	–	0.60	0.50	–
Saint Vincent and the Grenadines	–	–	0.34	0.30	–
Suriname	–	0.46	0.21	0.28	–
Trinidad and Tobago	0.69	0.31	0.59	0.64	–0.30
Uruguay	0.38	1.14	0.45	0.60	1.02
Venezuela, Bolivarian Rep. of	0.75	–1.27	0.38	0.33	0.47
SOUTH EAST ASIA REGION					
Bangladesh	0.70	0.29	0.69	0.77	0.74
Bhutan	1.26	–	0.59	0.82	0.98
India	0.42	0.40	0.50	0.60	0.07
Indonesia	0.89	0.67	0.80	0.91	0.86
Korea, Democratic People's Rep. of	–	–1.89	0.00	0.00	0.06
Maldives	0.68	–	0.87	0.93	–
Myanmar	0.17	1.00	0.26	0.29	0.51
Nepal	0.23	0.30	0.66	0.73	0.68
Sri Lanka	0.78	0.43	0.81	0.84	0.46
Thailand	1.42	0.60	0.63	0.65	1.04
Timor-Leste	–	–	0.90	0.97	–0.04
EUROPEAN REGION					
Albania	–2.20	–0.40	0.85	0.90	0.21
Armenia	–0.55	0.88	0.75	0.78	–0.13
Azerbaijan	–0.77	1.41	0.18	0.23	–4.36
Belarus	–	–	0.35	0.44	0.04
Bosnia and Herzegovina	–	0.00	0.42	0.48	–

	MDG 1: Eradicate extreme poverty and hunger		MDG 4: Reduce child mortality		MDG 5: Reduce maternal mortality
	Target 2: Halve, between 1990 and 2015, the proportion of people suffering from hunger		Target 5: Reduce by two thirds, between 1990 and 2015, the under 5 mortality rate		Target 6: Reduce by three quarters, between 1990 and 2015, the maternal mortality ratio
	Children under 5 moderately or severely underweight	Population undernourished	Infant mortality (0–1 year)	Children under 5 mortality	Maternal mortality
	1990–2006**	1990–2003**	1990–2005	1990–2005	1990–2000
Bulgaria	–	–0.25	0.30	0.25	–0.25
Croatia	–	1.13	0.68	0.63	–
Cyprus	–	–	0.90	0.88	–11.20
Czech Republic	–	–	1.09	1.04	0.53
Estonia	–	1.33	0.75	0.84	0.10
Georgia	–	1.41	0.07	0.06	0.04
Hungary	–	–	0.80	0.79	0.84
Israel	–	–	0.75	0.75	–1.14
Kazakhstan	–	–	–0.28	–0.24	–2.17
Kyrgyzstan	–	1.62	0.22	0.24	0.00
Latvia	–	0.00	0.54	0.58	–0.70
Lithuania	–	–	0.45	0.46	0.63
Macedonia, The Former Yugoslav Rep. of	–	1.07	0.82	0.83	–
Malta	–	–	0.67	0.68	–
Moldova, Rep. of	–	–2.40	0.80	0.90	0.53
Poland	–	–	1.03	0.92	0.63
Romania	–	–	0.61	0.58	0.74
Russian Federation	–	0.50	0.59	0.57	0.18
Serbia and Montenegro	–	–2.00	0.75	0.70	1.33
Slovakia	–	–1.00	0.63	0.64	–
Slovenia	–	0.00	0.94	0.90	–0.41
Tajikistan	–	–3.55	0.61	0.67	0.31
Turkey	0.65	–	0.92	0.97	0.81
Turkmenistan	–	0.67	–0.02	–0.11	0.58
Ukraine	–	–	0.47	0.52	0.32
Uzbekistan	–	–4.50	0.18	0.21	0.75
EASTERN MEDITERRANEAN REGION					
Afghanistan	–	–	0.03	0.02	–0.16
Bahrain	2.00	–	0.60	0.63	0.60
Djibouti	–0.09	–	0.42	0.28	–0.37
Egypt	0.96	0.50	0.95	1.02	0.67
Iran, Islamic Rep. of	–	0.00	0.64	0.75	0.49
Iraq	–0.17	–	–2.33	–2.25	0.26
Jordan	0.88	–1.50	0.50	0.53	0.97
Kuwait	1.64	1.58	0.54	0.47	0.78
Lebanon	–	–	0.23	0.28	0.67
Libyan Arab Jamahiriya	–	–	0.73	0.80	0.75
Morocco	–0.08	0.00	0.72	0.83	0.85
Oman	0.92	–	0.90	0.94	0.72
Pakistan	0.44	0.08	0.32	0.36	–0.63
Qatar	–	–	0.21	0.29	–
Saudi Arabia	–	0.00	0.60	0.61	1.10
Somalia	–	–	0.00	0.00	0.42
Sudan	–0.27	0.26	0.24	0.38	0.14
Syrian Arab Republic	0.50	0.40	0.90	0.99	0.15
Tunisia	1.02	–	0.77	0.81	0.39
United Arab Emirates	–	–	0.50	0.54	–1.44
Yemen	–0.85	–0.18	0.34	0.42	0.79

	MDG 1: Eradicate extreme poverty and hunger		MDG 4: Reduce child mortality		MDG 5: Reduce maternal mortality
	Target 2: Halve, between 1990 and 2015, the proportion of people suffering from hunger		Target 5: Reduce by two thirds, between 1990 and 2015, the under 5 mortality rate		Target 6: Reduce by three quarters, between 1990 and 2015, the maternal mortality ratio
	Children under 5 moderately or severely underweight	Population undernourished	Infant mortality (0–1 year)	Children under 5 mortality	Maternal mortality
	1990–2006**	1990–2003**	1990–2005	1990–2005	1990–2000
WESTERN PACIFIC REGION					
Brunei Darussalam	–	–	0.30	0.27	0.51
Cambodia	0.33	0.47	–0.34	–0.37	0.67
China	1.30	0.50	0.59	0.67	0.55
Cook Islands	–	–	0.52	0.56	–
Fiji	–	–	0.54	0.63	0.22
Kiribati	–	–	0.39	0.39	–
Korea, Rep. of	–	–	0.56	0.67	1.13
Lao People's Democratic Rep.	0.35	0.55	0.73	0.77	0.00
Malaysia	0.70	0.00	0.56	0.68	0.65
Marshall Islands	–	–	0.29	0.55	–
Micronesia, Federated States of	–	–	–0.46	–0.53	–
Mongolia	1.22	0.35	0.75	0.82	–0.92
Nauru	–	–	–	–	–
Niue	–	–	–	–	–
Palau	–	–	0.96	1.01	–
Papua New Guinea	–	–	0.39	0.40	0.90
Philippines	0.76	0.54	0.59	0.70	0.38
Samoa	–	–	0.60	0.63	1.33
Singapore	1.54	–	0.86	1.00	–0.67
Solomon Islands	–	–	0.55	0.81	–
Tonga	–	–	0.35	0.38	–
Tuvalu	–	–	0.34	0.48	–
Vanuatu	–	–	0.53	0.58	1.33
Viet Nam	0.81	0.90	0.87	0.96	0.25

Legend: Progressing and on track | Progressing but not on track | Going backwards

Indicator definitions

In determining progress towards meeting the MDGs, it was assumed that progress towards the target for each indicator is linear. The following approach has been used to determine whether countries are on track to meet the MDG targets and to quantify achievement to date.

1. Progress made for each indicator from 1990, the baseline year, to 2006 or the latest year for which data are available** (2006 – 1990).
2. Progress required to achieve the target for each indicator by 2015, using 1990 as the baseline year (2015 – 1990).
3. Divide (1) by (2): $\frac{(2006 - 1990)}{(2015 - 1990)}$

The results were interpreted as follows:

- A negative result indicates that the country is going backwards.
- A zero result indicates no progress has been made.
- A result between 0 and 0.5 indicates that the country is progressing, but not fast enough to achieve the target by 2015.
- A result greater than 0.5 indicates that the country is on track.
- A result of 1 indicates that the target has been achieved.

For the hunger indicators (MDG 1, Target 2), the 2015 target was calculated by halving the 1990 baseline figure:

$$\text{Target for 2015} = \frac{\text{1990 baseline figure}}{2}$$

For the child mortality indicators (MDG 4), the 2015 target was calculated by dividing the 1990 baseline figure by 3:

$$\text{Target for 2015} = \frac{\text{1990 baseline figure}}{3}$$

For the maternal mortality rate indicators (MDG 5), the 2015 target was calculated by dividing the 1990 baseline figure by 4:

$$\text{Target for 2015} = \frac{\text{1990 baseline figure}}{4}$$

The 1990 baseline data for the underweight indicator is based on estimates from the *Journal of the American Medical Association*, vol. 291, no. 21, 2 June 2004.

Note

A year span followed by two asterisks indicates progress made for each indicator from the baseline year to the latest year for which data are available.

Part VI Annexes

Abbreviations and acronyms

Glossary

Bibliography

Text notes

Costing the essential solutions

Methodology for maps

Abbreviations and acronyms

AIDS	acquired immune deficiency syndrome
ANC	antenatal clinic
ARI	acute respiratory infection
ART	anti-retroviral therapy
ARV	anti-retroviral
BMI	body mass index
CSB	corn–soya blend
CTC	community-based therapeutic care
CRED	Centre for Research on the Epidemiology of Disasters
DAC	Development Assistance Committee
DHS	demographic and health surveys
DOTS	directly observed treatment, short-course
DPT3	diphtheria, tetanus and whooping cough
DRC	Democratic Republic of the Congo
ECLAC	Economic Commission for Latin America and the Caribbean
EU	European Union
FAO	Food and Agriculture Organization of the United Nations
FBF	fortified blended food
GDP	gross domestic product
GNI	gross national income
HAART	highly active anti-retroviral therapy
HIV	human immunodeficiency virus
IDP	internally displaced person
IDB	Inter-American Development Bank
JUNJI	*Junta Nacional de Jardines Infantiles* (Chilean National Nursery School Council Programme)
LAC	Latin America and the Caribbean Region
LE	life expectancy
LIFDC	low-income food-deficit country
MCHN	mother-and-child health and nutrition
MDG	Millennium Development Goal
MDR	multi-drug resistant
MTCT	mother-to-child transmission
NGO	non-governmental organization
NHDP	National Health Development Plan
ODA	official development assistance
OECD	Organisation for Economic Co-operation and Development
OIE	World Organization for Animal Health (FAO)
ORS	oral rehydration salts
ORT	oral rehydration therapy
PLHIV	people living with HIV
SAM	severe acute malnutrition
SCN	Standing Committee on Nutrition (also UNSCN)
SFP	school feeding programme
TB	tuberculosis
UNAIDS	Joint United Nations Programme on HIV/AIDS
UNDAF	United Nations Development Assistance Framework
UNDP	United Nations Development Programme
UNEP	United Nations Environment Programme
UNESCO	United Nations Educational, Scientific and Cultural Organization
UNFPA	United Nations Population Fund
UNHCR	Office of the United Nations High Commissioner for Refugees
UN-HABITAT	United Nations Centre for Human Settlements
UNICEF	United Nations Children's Fund
UNSCN	United Nations Standing Committee on Nutrition
USA	United States of America
USAID	United States Agency for International Development
USDA	United States Department of Agriculture
VAD	vitamin A deficiency
VAM	vulnerability analysis and mapping
VCT	voluntary counselling and testing
WFP	World Food Programme
WHO	World Health Organization
WMO	World Meteorological Organization
WSB	wheat–soya blend
YLL	years of life lost

Glossary

Adolescence
The life stage between 10 and 19 years of age. Adolescents are particularly vulnerable to hunger because of the combined effects of increased physical activity, poor eating habits and/or lack of access to nutritious food, and unequal access to safe and improved livelihoods.

Anti-retroviral therapy (ART)
The administration of at least three different anti-retroviral drugs to suppress the replication of the human immunodeficiency virus. Treatment with this combination of drugs is also known as highly active anti-retroviral therapy. ART is not a cure. It must be taken for life and is delivered as part of a comprehensive care package that includes voluntary counselling and testing; the diagnosis and treatment of sexually transmitted diseases, TB and opportunistic infections; the prevention of mother-to-child transmission; and the treatment of pregnant women.

Avian influenza
An infectious disease caused by type-A strains of the influenza virus. The disease occurs worldwide. All birds are thought to be susceptible to infection with avian influenza viruses; many wild species carry these viruses with no apparent symptoms.

Body mass index
An indicator used to assess the nutritional status of adults and older children. It is derived by dividing the weight of an individual in kilograms by the square of the height measured in metres:

$$\frac{\text{Weight}}{\text{Height}^2}$$

Overweight adult: BMI >25
Underweight adult: BMI <18.5

Burden of disease
The significance of disease for society beyond the immediate cost of treatment, measured in years of life lost to ill health as the difference between total life expectancy and disability-adjusted life expectancy.

Care
Time, attention and support given by households and communities to meet the physical, mental and social needs of children and other household members.

Crude birth rate
Number of births per 1,000 population.

Crude death rate
Number of deaths per 1,000 population.

Diarrhoea
A symptom of infection caused by bacterial, viral and parasitic organisms, most of which can be spread by contaminated water. It is usually a harmless condition of short duration, but severe acute diarrhoea can lead to the loss of large amounts of fluids, dehydration and death.

Disease
Any deviation from or interruption of the normal structure or function of any body part, organ or system that is manifested by symptoms.

- *Chronic disease*
  A disease that has one or more of the following characteristics: it is permanent; leaves residual disability; is caused by irreversible pathological alteration; requires special training of the patient for rehabilitation; or may be expected to require a long period of supervision, observation or care.
- *Clinical disease*
  A disease with clinical signs and symptoms that is recognizable; distinct from a sub-clinical illness without clinical manifestations. Diabetes, for example, can be sub-clinical in someone before emerging as a clinical disease.
- *Deficiency disease*
  A condition caused by dietary or metabolic deficiency; includes diseases caused by an insufficient supply of essential nutrients.
- *Infectious disease*
  A disease caused by a bacterium, a virus, a fungus, a parasite or a prion. The infectious agent can enter the body in food, drink, air or bodily fluids. It then reproduces rapidly and subsequently infects other people. Some infectious diseases are benign; others are deadly and sometimes incurable.
- *Non-communicable disease*
  A disease that is not due to disease-causing organisms. It includes genetic diseases, such as Down's syndrome, haemophilia and those that are related to lifestyle or environment, such as cardiovascular disease and skin cancer.

Disease occurrence
Prevalence: The proportion of a population affected by a disease at a given time.
Incidence: Number of new cases of a disease in a population during a specified period.

DOTS (directly observed treatment, short-course)
The internationally recommended TB control strategy addresses challenges facing TB control by providing access to TB treatment and care; includes TB/HIV and MDR-TB patients.

Early childhood
As used in this report, the period from birth to 5 years of age.

Elderly
Age 60 and older. Traditional African definitions of an elder or elderly person correlate with age 50–65 years, depending on the region.

Food access (at household level)
A household's ability to acquire regular adequate amounts of food through a combination of home production and stocks, purchases, barter, gifts, borrowing or food assistance.

Food availability
The amount of food that is present in a country or area through all forms of domestic production, commercial imports and food assistance.

Food insecurity
Food insecurity, or the absence of food security, is a state that implies either hunger resulting from problems with availability, access and use or vulnerability to hunger in the future.

Food security
A condition that exists when all people, at all times, are free from hunger. The concept of food security provides insights into the causes of hunger. Food security has four parts:

- availability (the supply of food in an area);
- access (a household's ability to obtain that food);
- utilization (a person's ability to select, take-in and absorb the nutrients in the food); and
- vulnerability (the physical, environmental, economic, social and health risks that may affect availability, access and use).

Food utilization
This refers to the selection and intake of food and the absorption of nutrients into the body.

Gross domestic product
Sum of gross value added by all resident producers in the economy, including distributive trades and transport, plus any product taxes and minus any subsidies not included in the value of the products.

Gross national income
The total value of goods and services produced in a country (its GDP together with income received from other countries, mainly interest and dividends, minus similar payments made to other countries).

Growth faltering
Failure of a young child to grow to potential after birth, which is a physical indication that a child is not receiving the necessary micronutrients and macronutrients.

Health
A state of complete physical, mental and social well-being, not merely the absence of disease or infirmity. Health permits a socially and economically productive life.

Human immunodeficiency virus (HIV)
A virus that weakens the body's defences against disease. HIV makes the body vulnerable to a number of potentially life-threatening infections and cancers; it can be transmitted from one person to another. HIV causes a progressive decline in immunity that can lead to acquired immunodeficiency syndrome (AIDS). HIV/AIDS is used as a general term to refer to people who are infected with HIV who may or may not have the clinical diagnosis of AIDS. The more the immune system has been damaged, the greater the risk of death from opportunistic infections.

Hunger
A condition in which people lack the required nutrients, both macro (energy and protein) and micro (vitamins and minerals), for fully productive, active and healthy lives. Hunger can be a short-term phenomenon, or a longer-term chronic problem. It can have a range of effects from mild to severe. It can result from people not taking in sufficient nutrients or their bodies not being able to absorb the required nutrients. It can also result from poor food and childcare practices.

Infant mortality rate
Number of deaths per 1,000 live births up to, but not including, 1 year of age.

Life expectancy
The number of years a newborn baby would live if subjected to the mortality risks prevailing for each age group in the population.

Livelihoods
All capabilities, assets (including both material and social resources) and activities required for a means of living.

Low birthweight
A birthweight under 2,500 grams.

Macronutrients
These nutrients include carbohydrates, protein and fat. They form the bulk of the diet and provide all energy needs.

Malaria
A disease caused by a parasite called *Plasmodium*, which is transmitted by the bites of infected mosquitoes. The parasites multiply in the liver and then infect red blood cells. Symptoms of malaria include fever, headache, chills and vomiting between 10 and 15 days after the mosquito bite. If not treated, malaria can quickly become life-threatening by disrupting the blood supply to vital organs.

Malnutrition
A physical condition in which people experience either nutritional deficiencies (undernutrition) or an excess of certain nutrients (overnutrition).

- *Acute malnutrition*
  Protein-energy malnutrition caused by a recent and severe lack of food intake or disease that has led to substantial weight loss or nutritional oedema. There are different degrees/stages of acute malnutrition, which are often categorized as: severe malnutrition, moderate malnutrition and global acute malnutrition.
- *Global acute malnutrition*
  The percentage of children whose weight-for-height measurements fall below the cut-off of –2 standard deviations (or <–80 percent median) and/or who suffer from oedematous malnutrition. Note that GAM is sometimes referred to as "total" malnutrition.
- *Moderate malnutrition*
  The percentage of children whose weight-for-height measurements fall below the cut-off of –3 to –2 standard deviations (or 70 to 80 percent below the median).
- *Protein-energy malnutrition*
  A condition most commonly affecting children between the ages of 6 months and 5 years caused by lack of food or from infections that cause loss of appetite while increasing the body's nutrient requirements and losses. *Kwashiorkor* usually affects children aged 1–4 years, although it also occurs in older children and adults. The main sign is oedema. Because of oedema, children with *kwashiorkor* may look "fat" so that their parents regard them as well fed. *Marasmus* results from prolonged starvation or hunger; it may also result from chronic or recurring infections with marginal food intake. The main sign is a severe wasting.
- *Severe malnutrition*
  The percentage of children whose corresponding weight-for-height measurements fall below the cut-off of –3 standard deviations (or <70 percent below the median).

Maternal mortality rate
Number of deaths of women from pregnancy-related causes per 100,000 live births.

Maternal morbidity
All complications of pregnancy, delivery and abortion. It is noteworthy that the major portion of maternal ill health is found after pregnancy.

Micronutrients
These nutrients include all vitamins and minerals that in small amounts are essential for life.

Morbidity
The incidence or prevalence of disease in a given population.

Mortality
The number of deaths in a population at risk during a specified period.

Neonatal mortality rate
Number of deaths per 1,000 live births in the first 28 days after birth.

Obesity
A condition describing excess bodyweight in the form of fat. Morbid obesity is defined as being 100 pounds or nearly 50 kilograms overweight or having a BMI of 40 or above. Obesity is associated with many illnesses and is directly related to increased mortality and lower life expectancy.

Perinatal morbidity
Refers to complications of the foetus/infant and the mother. The road to perinatal health for the young mother is threatened by complications potentially affecting both the mother and the foetus; either of the two may die from complications originating in the woman's early years.

Perinatal mortality
The total number of foetal deaths from 22 weeks of pregnancy to the seventh day of life.

Post-neonatal mortality rate
Number of deaths per 1,000 live births during the first 28 days after birth and up to, but not including, 1 year of age, recorded in a given period.

Stunting
An indicator of chronic malnutrition calculated by comparing the height-for-age of a child with a reference population of well-nourished and healthy children. The prevalence of stunting reflects the long-term nutritional situation of a population.

Tuberculosis (TB)
A contagious disease. Like the common cold, it spreads through the air. Only people who are sick with TB in their lungs are infectious. A person needs only to inhale a small number of these TB germs, known as bacilli, to be infected.

Under 5 mortality rate
Number of deaths per 1,000 live births between birth and 5 years of age.

Undernutrition
The physical manifestation of hunger that results from serious deficiencies in one or a number of macronutrients and micronutrients. The deficiencies impede a person from maintaining growth, pregnancy, lactation, physical work, cognitive functions, and resisting and recovering from disease.

Undernourishment
The condition of people whose dietary energy consumption is continuously below a minimum requirement for fully productive, active and healthy lives. It is determined using a proxy indicator that estimates whether the food available in a country is sufficient to meet the energy (but not the protein, vitamins and minerals) requirements of the population. Unlike undernutrition, the indicator does not measure an actual outcome.

Vector-borne diseases
Malaria, dengue and dengue haemorrhagic fever, yellow fever, and West Nile fever are indirectly caused by flooding leading to standing water in which mosquitoes breed.

Vulnerability
The presence of factors that place people at risk of becoming hungry.

Wasting
An indicator of acute undernutrition that reflects a recent and severe process that has led to substantial weight loss. This is usually the result of starvation or disease, and is strongly related to mortality. It is calculated by comparing weight-for-height with a reference population of well-nourished people.

Z-score
A measure that shows where an individual value lies relative to a reference population. It is the deviation of an individual's value from the median value of a reference population, divided by the standard deviation of the reference population.

Bibliography

Part I

Australian Institute of Health and Welfare. 2007. *What Influences the Life Expectancy of a Population?* Posted at: http://www.aihw.gov.au/mortality/data/life_expectancy.cfm

Barker, D.J.P. 1998. *Mothers, Babies and Health in Later Life*. Edinburgh, UK, Churchill Livingstone.

Barnes, S. 2007. This month balance profiles: Jeanne Moe, Director of Project Archaeology. *Bozeman Daily Chronicle*, 2 January: 23.

Bermudez, O. & Dwyer, J. 1999. Identifying elders at risk of malnutrition: A universal challenge. ACC/Sub-Committee on Nutrition. *SCN News*, 19 December 1999.

Blum, R.W. 1991. Global trends in adolescent health. *Journal of the American Medical Association*, 265(20): 2711–2719.

Center for Disease Control. 2004. Mean body weight, height and body mass index 1960–2002. *Advance Note*, 347:1–18. Posted at: http://www.cdc.gov/nchs/data/ad/ad347.pdf

Cohen, M.N. & Armelagos, G.J. 1984. Paleopathology at the origins of agriculture. *In* W. Nicholson. 1999. *Longevity and Health in Ancient Paleolithic vs. Neolithic Peoples: Not What You May Have Been Told*. Posted at: http://www.beyondveg.com/nicholson-w/angel-1984/angel-1984-1a.shtml

Center for Research on the Epidemiology of Disasters. 2007a. Disaster data: A balanced perspective. *CRED Crunch* (8), March 2007. *Université Catholique de Louvain*. Posted at: http://www.em-dat.net

___ 2007b. *Emergency Disasters Data Base – Trends*. Posted at: http://www.em-dat.net/disasters/profiles.php

Disease Control Priorities Project. 2007. *Achieving the Millennium Development Goals for Health: So Far, Progress is Mixed – Can We Reach our Targets*? Washington DC, The World Bank Group. Posted at: http://www.dcp2.org/file/67/DCPP%20-%20MDGs.pdf

Engle, P. 1999. The role of caring practices and resources for care in child survival, growth and development: South and Southeast Asia. *Asian Development Review*, 17(1 and 2):132–167. Manila, Asian Development Bank.

Ending Child Hunger and Undernutrition Initiative. 2006. *Global Framework for Action*. Revised Draft. Rome, WFP and New York, UNICEF.

Food and Agriculture Organization of the United Nations. 2001. *The State of Food Insecurity in the World 2001*. Rome, FAO.

___ 2003. *The State of Food Insecurity in the World 2003*. Rome, FAO.

___ 2005. *The State of Food Insecurity in the World 2005*. Rome, FAO.

___ 2006a. *The State of Food Insecurity in the World 2006*. Rome, FAO.

___ 2006b. *The State of Food and Agriculture in the World: Food Aid for Food Security?* Rome, FAO.

Gautam, K.C. 2006. Nutrition: A life-cycle approach to support the Millennium Development Goals. *In* Standing Committee on Nutrition. 2006. *Tackling the Double Burden of Malnutrition: A Global Agenda. SCN News* No. 32.

Gillespie, S. 2001. Empowering women to achieve food security: Health and nutrition. *2020 Focus,* 6(08). Washington DC, IFPRI. Posted at: http://www.ifpri.org/2020/focus/focus06/focus06_08.asp

Global Health Council. 2007. *Women's Health: Maternal Health*. Posted at: http://www.globalhealth.org/view_top.php3?id=225

Gordon, B., Mackay, R. & Rehfuess, E. 2004. *Inheriting the World: The Atlas of Children's Health and the Environment*. Posted at: http://www.who.int/ceh/publications/en/atlas.pdf

Gwatkin, D.R. 2005. How much would the poor gain from faster progress towards the Millennium Development Goals for health? *Lancet*, 365: 813–817. Posted at: http://www.lancet.com

Holben, D.H. 2005. *The Concept and Definition of Hunger and Its Relationship to Food Insecurity.* Athens, OH, Ohio University. Posted at: http://www7.nationalacademies.org/cnstat/Concept_and_Definition_of_Hunger_Paper.pdf

Inter-American Development Bank. 1999. *Facing Up to Inequality in Latin America*. Report on Economic and Social Progress in Latin America, 1998–1999. Posted at: http://www.iadb.org/exr/english/PRESS_PUBS/ipintr.htm

Ismail, S. & Manandhar, M. 1999. *Better Nutrition for Older People – Assessment and Action.* London, London School of Hygiene and Tropical Medicine, and Help Age International.

Krishna, A., Kristjanson, P., Odero, A. & Nindo, W. 2004. Escaping poverty and becoming poor in 20 Kenyan villages. *Journal of Human Development*, 5(2).

Lindstrand, A., Bergström, S., Rosling, H., Rubenson, B., Stenson, B. & Tylleskar, T. 2006. *Global Health: An Introductory Textbook*. Copenhagen, Narayana Press.

Margallo, S. 2005. Addressing gender in conflict and post-conflict situations in the Phillipines. *In* F. Baingana, I. Bannon and R. Thomas. *Mental Health and Conflicts: Conceptual Framework and Approaches. Health, Nutrition and Population (HNP) Discussion Paper,* February 2005. Washington DC, World Bank.

Martorell, R. & Habitch, J-P. 1986. Growth in early childhood in developing countries. *In* T. P. Schultz. 2003. *Human Capital, Schooling and Health Returns. Economic Growth Center Discussion Paper Series* No. 853. Posted at: http://www.econ.yale.edu/growth_pdf/cdp853.pdf

Mason, J.B. 2002. Supplement: History of food and nutrition in emergency relief. Lessons on nutrition of displaced people. *Journal of Nutrition*, 132: 2096S–2103S.

McCormick, M.C. 1985. The contribution of low birthweight to infant mortality and childhood morbidity. *New England Journal of Medicine,* 312: 82–90.

Millman, S. & Kates, R.W. 1990. Toward understanding hunger. *In* L.F. Newman et al. eds. 1990. *Hunger in History: Food Shortage, Poverty, and Deprivation.* Cambridge, MA, Blackwell.

Mills, A. & Shillcutt, S. 2004. *Communicable Diseases. Copenhagen Consensus Challenge Paper.* Posted at: http://www.copenhagenconsensus.com/Default.aspx?ID=158

Milton, K. 2000. Hunter-gatherer diets: A different perspective. *The American Journal of Clinical Nutrition*, 71(3): 665–667. Posted at: http://www.ajcn.org/

Nestel, P. 2000. *Strategies, Policies, and Programs to Improve the Nutrition of Women and Girls*. Draft. Washington DC. Food and Nutrition Technical Assistance (FANTA) Project, Academy for Educational Development. Posted at: http://www.fantaproject.org/downloads/pdfs/StrategiesPoliciesPrograms_Nestel.pdf

Newman, L.F., Boegehold, A., Heeley, D., Kates, R.W. & Raaflaub, K. 1990. Agricultural intensification, urbanization and hierarchy. *In* L.F. Newman et al. eds. 1990. *Hunger in History: Food Shortage, Poverty, and Deprivation*. 1990. Cambridge, MA, Blackwell.

Organisation for Economic Co-operation and Development. 2007. *Statistics Portal.* Paris. Posted at: http://www.oecd.org/statsportal/0,3352,en_2825_293564_1_1_1_1_1,00.html

Pan American Health Organization. 2007. *Health of the Indigenous Peoples Initiative, Strategic Directions and Plan of Action 2003–2007.* Posted at: http://www.paho.org/English/AD/THS/OS/Plan2003-2007-eng.doc

Richards, A.I. 2003. *Hunger and Work in a Savage Tribe: A Functional Study of Nutrition among the Southern Bantu*. London, Routledge.

Russell, S.A. 2005. *Hunger: An Unnatural History.* New York, Basic Books Inc.

Salleh, M.N. 2001. Globeglance: We can have enough food. *United Nations Chronicle,* online edition. Posted at: http://www.un.org/Pubs/chronicle/2001/issue3/0103p44.html

Schroeder, D. 2001. Malnutrition. *In* R.D. Semba and M.W. Bloem eds. 2001. *Nutrition and Health in Developing Countries*. Totowa, NJ, Humana Press.

Schultz, T.P. 2002. Wage gains associated with height as a form of health human capital. *American Economic Review* 92(2): 349–353. *In* T.P. Schultz. 2003. *Human Capital, Schooling and Health Returns. Economic Growth Center Discussion Paper Series*, No. 853. Posted at: http://www.econ.yale.edu/growth_pdf/cdp853.pdf

Scrimshaw, N.S., Munoz, J.A., Tandon, O.B. & Guzman, M.A. 1959. Growth and development of Central American children II: The effect of oral administration of vitamin B12 to rural children of pre-school and school age. *American Journal of Clinical Nutrition*, 7(2): 180–184.

Semba, R.D. 2001. Nutrition and development: A historical perspective. *In* R.D Semba and M.W. Bloem eds. 2001. *Nutrition and Health in Developing Countries.* Totowa, NJ, Humana Press.

Shaw, G.B. (1856–1950). *Back to Methuselah.* Posted at: http://en.proverbia.net/citasautor.asp?autor=16652

Siega-Riz, A.M., Adair, L.S. & Hobel, C.J. 1994. Institute of Medicine maternal weight gain recommendations and pregnancy outcome in a predominantly Hispanic population. *Obstetrics and Gynecology*, 84: 565–573.

Spurr, G.B., Bares-Nieto, M. & Maksud, M.G. 1977. Productivity and maximal oxygen consumption in sugar cane cutters. *American Journal of Clinical Nutrition*, 30: 316–321.

United Nations. 2007. *The Millennium Development Goals Report 2007*. New York. Posted at: http://www.un.org/millenniumgoals/pdf/mdg2007.pdf

Joint United Nations Programme on HIV/AIDS. 2006. *AIDS Epidemic Update: December 2006.* Geneva. Posted at: http://data.unaids.org/pub/EpiReport/2006/2006_EpiUpdate_en.pdf

United Nations Children's Fund. 2000. Evaluation Report – SEN 2002/001: *Impacts Socio-Economiques du VIH/SIDA sur les Enfants: Le Cas du Sénégal.* Posted at: http://www.unicef.org/evaldatabase/index_14243.html
___ 2004. *Children on the Brink, 2004.* New York, UNICEF.
___ 2006a. *Progress for Children. A Child Survival Report Card on Nutrition.* New York, UNICEF.
___ 2006b. *The State of the World's Children 2007 – Women and Children: The Double Dividend of Gender Equality.* New York, UNICEF.
___ 2007. *Malaria.* New York. Posted at: http://www.unicef.org/health/index_malaria.html

United Nations Development Programme. 2005. Sub-Saharan Africa – the human costs of the 2015 "business-as-usual" scenario. *Human Development Report 2005.* Posted at: http://hdr.undp.org/docs/events/Berlin/Background_paper.pdf

United Nations Millennium Project – Hunger Task Force. 2005. *Halving Hunger: It Can Be Done.* Lead authors: P. Sanchez, M.S. Swaminathan, P. Dobie and N. Yuksel. New York. Posted at: http://www.unmillenniumproject.org/documents/HTF-SumVers_FINAL.pdf

United Nations Population Division. 2007. *United Nations Population Information Network (POPIN).* New York. Posted at: http://www.un.org/popin/

United Nations Population Fund. 2005. *The State of the World Population – the Promise of Equality: Gender Equity, Reproductive Health and the Millennium Development Goals.* New York, UNFPA.

United Nations Statistics Division. 2007. *Millennium Development Goals Indicators.* New York. Posted at: http://mdgs.un.org/unsd/mdg/Defaults.aspx

Waldman, R. 2005. Public health in war: Pursuing the impossible. *International Health*, 27(1): 1–5.

Webb, P. & Rogers, B. 2003. Addressing the "In" in Food Insecurity. *USAID Occasional Paper No. 1.* Posted at http://www.dec.org/pdf_docs/PNACS926.pdf.

Whitehead, J.W. 1983. *The Stealing of America.* Westchester, NY, Crossway Books.

World Bank. 2006. *Repositioning Nutrition as Central to Development: A Strategy for Large-Scale Action. Washington DC.* Posted at: http://siteresources.worldbank.org/NUTRITION/Resources/281846-1131636806329/NutritionStrategy.pdf
___ 2007. *Global Monitoring Report 2007: Millennium Development Goals. Confronting the Challenges of Gender Equality and Fragile States.* Washington DC, World Bank.

World Food Programme. 2002. *VAM Standard Analytical Framework Guideline. Role and Objectives of VAM Activities to Support WFP Food-Oriented Interventions.* Rome, WFP.
___ 2004. *Nutrition in Emergencies: WFP Experiences and Challenges.* WFP/EB.A/2004/5-A/3. Rome, WFP.
___ 2005. *Hunger Inequalities in Latin America and the Caribbean: Reworking Social Programmes in the Face of Globalization.* Panama, WFP.
___ 2006a. WFP Emergency Needs Assessment Report: Pre-Crisis Information. *Executive Brief: Liberia*, 23 September 2006. Rome, WFP.
___ 2006b. *World Hunger Series 2006: Hunger and Learning.* Rome, WFP and Palo Alto, CA, Stanford University.
___ 2007. Kenya's refugee camps. *WFP Latest News – In Depth.* Posted at: http://www.wfp.org/newsroom/in_depth/africa/kenya/050307_kenya_kakuma.asp?section=2&sub_section=2

World Health Organization. 1997. *New Report Confirms Global Spread of Drug-Resistant Tuberculosis. WHO Press Release* 74, 22 October.
___ 2005. *Health and the Millennium Development Goals.* Geneva. Posted at: http://www.who.int/mdg/publications/mdg_report/en/
___ 2006. *The World Health Report: Working Together for Health.* Geneva, WHO.
___ 2007a. *Core Health Indicators – World Health Statistics.* Geneva. Posted at: http://www.who.int/en
___ 2007b. *Towards Universal Access: Scaling Up Priority HIV/AIDS Interventions in the Health Sector.* Progress Report, April 2007. Geneva. Posted at: http://www.who.int/hiv/mediacentre/universal_access_progress_report_en.pdf
___ 2007c. *Tuberculosis (TB): Frequently Asked Questions about TB and HIV.* Geneva. Posted at: http://www.who.int/tb/hiv/faq/en/
___ 2007d. *WHO Report 2007. Global Tuberculosis Control: Surveillance, Planning and Financing.* Geneva. Posted at: http://www.who.int/tb/publications/global_report/en/
___ 2007e. *World Health Statistics 2007.* Geneva. Posted at: http://www.who.int/whosis/whostat2007/en

World Health Organization, United Nations Population Fund & United Nations Children's Fund. 1995. *Action for Joint Adolescent Health: Towards a Common Agenda. Recommendations from a Joint Study Group.* Posted at: http://www.who.int/child-adolescent-health/New_Publications/ADH/WHO_FRH_ADH_97.9_en.pdf

Yach, D. 1998. Health and illness: The definition of the World Health Organization. Posted at: http://www.medezin-ethik.ch

Intermezzo 1: "An overview of Micronutrient Deficiencies".

Basu, S., Sengupta, B. & Roy Paladhi, P.K. 2003. Single megadose vitamin A supplementation of Indian mothers and morbidity in breastfed young infants. *Postgraduate Medical Journal*, 79: 397–402.

Caulfield, L.E. & Black, R.E. 2004. Zinc deficiency. *In* M. Ezzati, A.D. Lopez, A. Rogers and C.L.J. Murray. 2004. *Comparative Quantification of Health Risks*. Geneva, WHO.

Dugdale, M. 2001. Anaemia. *Obstetrics and Gynecology Clinics of North America*, 28(2): 363–381.

Green, N.S. 2002. Folic acid supplementation and prevention of birth defects. *Journal of Nutrition* 132 (8 Supplement): 2356S–2360S.

Gupta, H. & Gupta, P. 2004. Neural tube defects and folic acid. *Indian Pediatrics*, 41: 577.

Johnson, W.G., Scholl, T.O., Spychala, J.R., Buyske, S., Stenroos, E.S. & Chen, X. 2005. Common dihydrofolate reductase 19-base pair deletion allele: A novel risk factor for preterm delivery. *American Journal of Clinical Nutrition*, 81(3): 664–668.

Rice, A.L., West Jr., K.P. & Black, R.E. 2004. Vitamin A Deficiency. *In* M. Ezzati, A.D. Lopez, A. Rogers and C.L.J. Murray. 2004. *Comparative Quantification of Health Risks*. Geneva, WHO.

Sanghvi, T., Ross, J. & Heyman, H. 2007. Why is reducing vitamin and mineral deficiencies critical for development? The links between VMDs and survival, health, education, and productivity. *Food and Nutrition Bulletin*, 28(1) (Supplement): S170.

Stoltzfus, R.J., Mullany, L. & Black, R.E. 2004. Iron deficiency anaemia. *In* M. Ezzati, A.D. Lopez, A. Rogers and C.L.J. Murray. 2004. *Comparative Quantification of Health Risks*. Geneva, WHO.

World Health Organization. 2002. *The World Health Report: Reducing Risks, Promoting Healthy Life*. Geneva. Posted at: http://www.who.int/whr/2002/en/

Intermezzo 2: Joan Holmes, President of the Hunger Project. 2007. "Women and Ending Hunger: The Inextricable Link".

Food and Agriculture Organization of the United Nations. 2006. *The State of Food Insecurity in the World 2006*. Rome, FAO.

Food Corporation of India. 2007. Stock Management. New Delhi, Government of India. Posted at: http://fciweb.nic.in/stock_management/stock_management.htm

Osmani, S. & Sen, A. 2003. The hidden penalties of gender inequality: Fetal origins of ill health. *Economics and Human Biology*, 1: 105–121.

Ramalingaswami, V., Jonsson, U. & Rohde, J. 1996. The Asian enigma. *In The Progress of Nations*. New York, UNICEF.

Smith, L. & Haddad, L. 2000. Overcoming child malnutrition in developing countries: Past achievements and future choices. *IFPRI Discussion Paper* No. 30. Washington DC, IFPRI.

World Health Organization. 2002. *Death and DALY Estimates for 2002 by Cause for WHO Member States*. Posted at: http://www.who.int/entity/healthinfo/statistics/en

Intermezzo 3: Patrick Webb, Dean for Academic Affairs, Friedman School of Nutrition Science and Policy, Tufts University and Andrew Thorne-Lyman, Public Health Nutrition Officer, WFP. 2007. "Hunger and Disease in Crisis Situations".

Excerpt from a paper prepared for the UNU-WIDER project on Hunger and Food Security: New Challenges and New Opportunities, directed by Basudeb Guha-Khasnobis.

Center for Research on Epidemiology of Disasters. 2005. *EM_DAT: The OFDA/CRED International Disaster Database*. Belgium. Posted at: http://www.em-dat.net

Iliffe, J. 1990. *Famine in Zimbabwe 1890–1960*. Harare, Mambo Press.

Sen, A. 1997. *Entitlement Perspectives of Hunger*. Paper presented at a WFP/UNU seminar, 31 May. Rome.

Toole, M.J. & Waldman, R.J. 1988. An analysis of mortality trends among refugee populations in Thailand, Somalia and Sudan. *Bulletin of World Health Organization*, 6: 237–247.

Webb, P. 2002. Emergency relief during Europe's famine of 1817 anticipated crisis-response mechanisms of today. *Journal of Nutrition*, 132(7): 2092S–2095S.

___ 2003. The under-resourcing of rights: Empty stomachs and other abuses of humanity. *New England Journal International and Comparative Law*, 9(1): 135–157.

Part II

Alaimo, K., Olson, C.M. & Frongillo Jr., E.A. 2001. Food insufficiency and American school-aged children's cognitive, academic and psychosocial development. *Pediatrics*, 108(1): 44–53.

Anderson, S.E., Cohen, P., Naumova, E.N. & Must, A. 2006. Association of depression and anxiety disorders with weight change in a prospective community-based study of children followed up into adulthood. *Archives of Paediatrics and Adolescent Medicine*, 160(3): 285–291.

Ash, N. 2005. *Earth Species Feel the Squeeze*. BBC News website, 21 May 2005. Posted at: http://news.bbc.co.uk/2/low/science/nature/4563499.stm

Assis, A.M., Barreto, M.L., Santos, L.M., Fiaccone, R. & Da Silva Gomes, G.S. 2005. Growth faltering in childhood related to diarrhoea: A longitudinal community based study. *European Journal of Clinical Nutrition*, 59(11): 1317–1323.

Barnes, S. 2007. This month balance profiles: Jeanne Moe, Director of Project Archaeology. *Bozeman Daily Chronicle*, 2 January: 23.

Bates, I., Fenton, C., Gruber, J., Lalloo, D., Lara, A.M., Squire, S.B., Theobald, S., Thomson, R. & Tolhurst, R. 2004. Vulnerability to malaria, tuberculosis and HIV/AIDS infection and disease. Part II: determinants operating at individual and institutional level. *Lancet Infectious Diseases,* 4(5): 267–277.

Briend, A., Hasan, K.Z., Aziz, K.M. & Hoque, B.A. 1989. Are diarrhoea control programmes likely to reduce childhood malnutrition? Observations from rural Bangladesh. *Lancet* 2 (8658): 319–322.

Bryce, J., Boschi-Pinto, C., Shibuya, K., Black, R.E. & WHO Child Health Epidemiology Reference Group. 2005. WHO estimates of the causes of death in children. *Lancet*, 365(9465): 1147–1152.

Cegielski, J.P. & McMurray, D.N. 2004. The relationship between malnutrition and tuberculosis: Evidence from studies in humans and experimental animals. *International Journal of Tuberculosis and Lung Disease*, 8(3): 286–298.

Chatterjee, B.D., Bhattacharyya, A.K. & Mandal, J.N. 1968. Serum proteins in kwashiorkor and marasmus. 2. Non-tuberculous cases and tuberculous cases. *Bulletin of the Calcutta School of Tropical Medicine*, 16(3): 73–74.

Checkley, W., Epstein, L.D., Gilman, R.H., Cabrera, L. & Black, R.E. 2003. Effects of acute diarrhoea on linear growth in Peruvian children. *American Journal of Epidemiology*, 157(2): 166–175.

Coutsoudis, A., Coovadia, H., Pillay, K. & Kuhn, L. 2001. Are HIV-infected women who breastfeed at increased risk of mortality? *AIDS,* 15(5): 653-655.

Deb, S.K. 1998. Acute respiratory disease survey in Tripura in case of children below five years of age. *Journal of Indian Medical Association,* 96(4): 111–116.

Delisle, H., Chandra-Mouli, V. & de Benoist, B. 2000. Should Adolescents be Specifically Targeted for Nutrition in Developing Countries? To Address Which Problems, and How? Posted at: http://www.who.int/child-adolescent-health/New_Publications/NUTRITION/Adolescent_nutrition_paper.pdf

Des Moines Declaration. 2004. *A Call for Accelerated Action in Agriculture, Food and Nutrition to End Poverty and Hunger*. World Food Prize Day, 16 October. Des Moines, IA, World Food Prize Organization. Posted at: http://www.worldfoodprize.org/assets/laureates/statements/dmdeclaration.pdf

Fawzi, W. & Mehta, S. 2007. Hunger and HIV/AIDS and tuberculosis (TB). Technical paper for the *World Hunger Series 2007*. Cambridge, MA, Harvard University.

Fawzi, W., Msamanga, G., Spiegelman, D. & Hunter D. J. 2005. Studies of vitamins and minerals and HIV transmission and disease progression. *Journal of Nutrition*, 135(4): 938–944.

Food and Agriculture Organization of the United Nations. 2003. *The State of Food Insecurity in the World 2003*. Rome, FAO.

___ 2006. Will the desert locusts strike again? *FAO Newsroom*. Posted at: http://www.fao.org/ag/locusts

Food and Agriculture Organization of the United Nations & World Organization for Animal Health. 2005. *Preparing for the Highly Pathogenic Avian Influenza*. Rome, FAO.

Godfrey, K.M. & Barker, D.J. 2001. Foetal programming and adult health. *Public Health and Nutrition*, 4(2B): 611–624.

Gortmaker, S.L., Must, A., Perrin, J.M., Sobol, A.M. & Dietz, W.H. 1993. Social and economic consequences of overweight in adolescence and young adulthood. *New England Journal of Medicine*, 329(14): 1008–1012.

Greenblott, K. 2007. Social protection in the era of HIV and AIDS. Examining the role of food-based interventions. *WFP Occasional Paper* No. 17. Rome, WFP.

Guha-Sapir, D., Hargitt, D. & Hayois, P. 2004. *Thirty Years of Natural Disasters 1974–2003: The Numbers. Presses Universitaires de Louvain*. Louvain, Belgium, CRED. Posted at: http://www.em-dat.net/documents/Publication/publication_2004_emdat.pdf

Harries, A.D., Nkhoma, W.A., Thompson, P.J., Nyangulu, D.S. & Wirima, J.J. 1988. Nutritional status in Malawian patients with pulmonary tuberculosis and response to chemotherapy. *European Journal of Clinical Nutrition*, 42(5): 445–450.

Harries, A.D., Thomas, J. & Chugh, K.S. 1985. Malnutrition in African patients with pulmonary tuberculosis. *Human Nutrition Clinical Nutrition Journal*, 39(5): 361–363.

Hewson, P.D. (Bono). 2005. This generation's moon shot. *Time Magazine*, 1 November 2005. Posted at: http://www.time.com/time/magazine/article/0,9171,1124333,00.html

Hippocrates. (460–380 BC). *Hippocratic Writings: On Airs, Waters and Places.* Translated by F. Adams. Posted at: http://www.4literature.net/Hippocrates/On_Airs_Waters_and_Places/

Intergovernmental Panel on Climate Change. 2007. Summary for policymakers. *In* M.L. Parry, O.F. Canziani, J.P. Palutikof, P.J. van der Linder and C.E. Hanson. 2007. *Climate Change 2007: Impacts, Adaptation and Vulnerability*. Contribution of Working Group II to the Fourth Assessment Report of the IPCC. Geneva. Posted at http://www.ipcc.ch/SPM13apr07.pdf

Kupka, R., Garland, M., Msamanga, G., Spiegelman, D., Hunter, D. & Fawzi, W. 2005. Selenium status, pregnancy outcomes and mother-to-child transmission of HIV-1. *Journal of Acquired Immune Deficiency Syndromes*, 39(2): 203–10.

Leyton, G.B. 1946. Effects of slow starvation. *Lancet*, 251(8): 73–79.

Lutter, C.K., Habicht, J-P., Rivera, J.A. & Martorell, R. 1992. The relationship between energy intake and diarrhoeal disease in their effects on child growth: Biological model, evidence, and implications for public health policy. *Food and Nutrition Bulletin*, 14(1): 36–42.

Macallan, D.C., McNurlan, M.A., Kurpad, A.V., de Souza, G., Shetty, P.S., Calder, A.G. & Griffin, G.E. 1998. Whole body protein metabolism in human pulmonary tuberculosis and undernutrition: Evidence for anabolic block in tuberculosis. *Clinical Science Journal*, 94(3): 321–331.

Megazzini, K., Washington, S., Sinkala, M., Lawson-Marriott, S., Stringer, E., Krebs, D., Levy, J., Chi, B., Cantrell, R., Zulu, I., Mulenga, L. & Stringer, J. 2006. *A Pilot Randomized Trial of Nutritional Supplementation in Food Insecure Patients Receiving Anti-Retroviral Therapy (ART) in Zambia*. Research paper presented at the XVI HIV/AIDS Conference. 16–18 August 2006. Toronto, Canada.

Michels, K.B. 2003. Early life predictors of chronic disease. *Journal of Women's Health*, 12(2): 157–161.

Mitnick, C., Bayona, J., Palacios, E., Shin, S., Furin, J., Alcántara, F., Sánchez, E., Sarria, M., Becerra, M., Fawzi, M.C., Kapiga, S., Neuberg, D., Maguire, J.H., Kim, J.Y. & Farmer, P. 2003. Community-based therapy for multi-drug resistant tuberculosis in Lima, Peru. *New England Journal of Medicine*, 348(2): 119–128.

Moore, S.R., Lima, A.A., Conaway, M.R., Schorling, J.B., Soares, A.M. & Guerrant, R.L. 2001. Early childhood diarrhoea and helminthiases associate with long-term linear growth faltering. *International Journal of Epidemiology*, 30(6): 1457–1464.

Nussenblatt, V. & Semba, R.D. 2002. Micronutrient malnutrition and the pathogenesis of malarial anaemia. *Acta Tropica*, 82: 321–337.

Onwubalili, J.K. 1988. Malnutrition among tuberculosis patients in Harrow, England. *European Journal of Clinical Nutrition*, 42(4): 363–366.

Paton, N.I., Castello-Branco, L.R., Jennings, G., Ortigao-de-Sampaio, M.B., Elia, M., Costa, S. & Griffin, G.E. 1999. Impact of tuberculosis on the body composition of HIV-infected men in Brazil. *Journal of Acquired Immune Deficiency Syndromes and Human Retrovirology*, 20(3): 265–271.

Paton, N.I., Chua, Y.K., Earnest, A. & Chee, C.B. 2004. Randomized controlled trial of nutritional supplementation in patients with newly diagnosed tuberculosis and wasting. *American Journal of Clinical Nutrition*, 80(2): 460–465.

Paton, N.I., Ng, Y-M., Chee, C.B., Persaud, C. & Jackson, A.A. 2003. Effects of tuberculosis and HIV infection on whole-body protein metabolism during feeding, measured by the [^{15}N] glycine method. *American Journal of Clinical Nutrition*, 78: 319–325.

Pelletier, D.L., Frongillo, E.A., Schroeder, D.G. & Habicht, J-P. 1995. The effects of malnutrition on child mortality in developing countries. *Bulletin of the World Health Organization*, 73(4): 443–448.

Pimentel, D. 1993. Climate changes and food supply. *Forum for Applied Research and Public Policy*, 8(4): 54–60. Ithaca, NY, Cornell University. Posted at: http://www.ciesin.columbia.edu/docs/004-138/004-138.html

Rahman, M.M., Mahalanabis, D., Alvarez, J.O., Wahed, M.A., Islam, M.A., Habte, D. & Khaled, M.A. 1996. Acute respiratory infections prevent improvement of vitamin A status in young infants supplemented with vitamin A. *Journal of Nutrition*, 126(3): 628–633.

Rice, A.L., West Jr., K.P. & Black, R.E. 2004. Vitamin A Deficiency. *In* M. Ezzati, A.D. Lopez, A. Rogers and C.L.J. Murray. 2004. *Comparative Quantification of Health Risks.* Geneva, WHO.

Roseboom, T.J., van der Meulen, J.H., Osmond, C., Barker, D.J., Ravelli, A.C. & Bleker, O.P. 2000a. Plasma lipid profiles in adults after prenatal exposure to the Dutch famine. *American Journal of Clinical Nutrition*, 72(5): 1101–1106.
___ 2000b. Coronary heart disease after prenatal exposure to the Dutch famine, 1944-45. *Heart*, 84(6): 595–598.
___ 2001. Adult survival after prenatal exposure to the Dutch famine, 1944–45. *Paediatric and Perinatal Epidemiology*, 15(3): 220–225.

Scalcini, M., Occenac, R., Manfreda, J. & Long, R. 1991. Pulmonary tuberculosis, human immunodeficiency virus type-1 and malnutrition. *Bulletin of the International Union against Tuberculosis and Lung Disease*, 66(1): 37–41.

Schneider, D. 2000. International trends in adolescent nutrition. *Social Science of Medicine*, 51(6): 955–967.

Schorling, J.B., McAuliffe, J.F., de Souza, M.A. & Guerrant, R.L. 1990. Malnutrition is associated with increased diarrhoea incidence and duration among children in an urban Brazilian slum. *International Journal of Epidemiology*, 19: 728–735.

Schroeder, D.G. 2001. Malnutrition. *In* R.D. Semba and M.W. Bloem eds. 2001. *Nutrition and Health in Developing Countries.* Totowa, NJ, Humana Press.

Scrimshaw, N. & San Giovanni, J.P. 1997. Synergism of nutrition, infection, and immunity: An overview. *American Journal of Clinical Nutrition*, 66(2): 464.

Scott, S. & Duncan, C.J. 2001. *Biology of Plagues: Evidence from Historical Populations.* Cambridge, UK, Cambridge University Press.

Shah, S., Whalen, C., Kotler, D.P., Mayanja, H., Namale, A., Melikian, G., Mugerwa, R. & Semba, R.D. 2001. Severity of human immunodeficiency virus infection is associated with decreased phase angle, fat mass and body cell mass in adults with pulmonary tuberculosis infection in Uganda. *Journal of Nutrition*, 131: 2843–2847.

Shankar, A.H. 2000. Nutritional modulation of malaria morbidity and mortality. *Journal of Infectious Diseases*, 182 (Supplement) 1: S37–S53.

Sorensen, T.I.A., Holst, C. & Stunkard, A.J. 1992. Childhood body mass index: Genetic and familial environmental influences assessed in a longitudinal adoption study. *International Journal of Obesity Related Metabolic Disorders*, 16: 705–714.

Standing Committee on Nutrition. 1990. Appropriate uses of anthropometric indices in children. *Nutrition Policy Discussion Paper*, No. 7.
___ 2004a. *Fifth Report on the World Nutrition Situation: Nutrition for Improved Development Outcomes.* Posted at: http://www.unsystem.org/scn/publications/AnnualMeeting/SCN31/SCN5Report.pdf
___ 2004b. Nutrition and the Millennium Development Goals. *SCN News*, No. 28.

Stephensen, C.B. 1999. Burden of Infection on Growth Failure. *Journal of Nutrition*, (129): 534–537.

Tomkins, A. & Watson, F. 1989. Malnutrition and infection – A review – Nutrition policy. *ACC/SCN State-of-the-Art Series. Discussion Paper* No. 5. Posted at: http://www.unsystem.org/SCN/archives/npp05/begin.htm

United Nations Children's Fund. 2002. *Facts for Life. With Advice on: Safe Motherhood, Breastfeeding, Child Development, Nutrition and Growth, Immunization, Diarrhoea, Malaria, HIV/AIDS and Much More…* New York. Posted at: http://www.unicef.org/ffl/pdf/factsforlife-en-full.pdf
___ 2006a. *Progress for Children. A Child Survival Report Card on Nutrition.* New York. Posted at: http://www.unicef.org/progressforchildren/
___ 2006b. *The State of the World's Children 2007 – Women and Children. The Double Dividend of Gender Equality.* New York, UNICEF.
___ 2007a. *Child Survival Fact Sheet: Water and Sanitation*. New York. Posted at: http://www.unicef.org/media/media_21423.html
___ 2007b. *Malaria*. New York. Posted at: http://www.unicef.org/health/index_malaria.html

United Nations Children's Fund & World Health Organization. 2006. *Pneumonia: The Forgotten Killer of Children.* New York, UNICEF and Geneva, WHO. Posted at: http://www.who.int/child-adolescent-health/New_Publications/CHILD_HEALTH/ISBN_92_806_4048_8.pdf

Joint United Nations Programme on HIV/AIDS. 2006. *Report on the Global AIDS Epidemic 2006.* Geneva. Posted at: http://www.unaids.org/en/HIV_data/2006GlobalReport/default.asp

United Nations Centre for Human Settlements. 2006. *State of the World's Cities – 2006/7.* Nairobi, UN-HABITAT and London, Earthscan.
___ 2007. *Global Urban Observatory.* Nairobi. Posted at: http://ww2.unhabitat.org/programmes/guo/statistics.asp

Villamor, E., Dreyfuss, M.L., Baylin, A., Msamanga, G. & Fawzi W. 2004. Weight loss during pregnancy is associated with adverse pregnancy outcomes among HIV-1 infected women. *Journal of Nutrition*, 134(6): 1424–1431.

Williams, C.D., Oxon, B.M. & Lond, H. 2003. Kwashiorkor. A nutritional disease of children associated with a maize diet. *Bulletin of the World Health Organization*, 81(12): 912–913.

World Food Programme. 2006. *WFP Emergency Needs Assessment Report/Pre-Crisis Information: Liberia Country assessment.* Rome, WFP.
___ 2007. *The Strategic Implications for WFP of the Emergence of Biofuels.* Rome, WFP. Mimeo.

World Health Organization. 2003. *Nutrient Requirements for People Living with HIV/AIDS.* Geneva, WHO.
___ 2005. *Preventing Chronic Diseases: A Vital Investment.* Geneva. Posted at: http://www.searo.who.int/en/Section1174/Section1459_10496.htm
___ 2006a. *Enriching Lives: Overcoming Under- and Over-Nutrition.* Global Programming Note 2006–2007. Geneva. Posted at: http://www.who.int/nmh/donorinfo/nutrition/nutrition_helvetica.pdf
___ 2006b. *Avian Influenza in Africa: Statement by the Director-General of WHO.* Geneva. Posted at: http://www.who.int/mediacentre/news/statements/2006/s03/en/index.html
___ 2007a. *WHO Report 2007. Global Tuberculosis Control: Surveillance, Planning and Financing.* Geneva. Posted at: http://www.who.int/tb/publications/global_report/en/
___ 2007b. *Networking for Policy Change: TB/HIV Advocacy Training Manual.* Geneva. Posted at: http://whqlibdoc.who.int/publications/2007/a90084_eng.pdf
___ 2007c. *Obesity and Overweight. Global Strategy on Diet, Physical Activity and Health.* Geneva. Posted at: http://www.who.int/dietphysicalactivity/publications/facts/obesity/en/
___ 2007d. *Core Health Indicators – World Health Statistics.* Geneva. Posted at: http://www.who.int/whosis/whostat2007/en/index.html
___ 2007e. *Cumulative Number of Confirmed Human Cases of Avian Influenza A/(H5N1) Reported to WHO.* Geneva. Posted at: http://www.who.int/csr/disease/avian_influenza/country/cases_table_2007_06_15/en/index.html

Zachariah, R., Spielmann, M.P., Harries, A.D. & Salaniponi, F.M. 2002. Moderate to severe malnutrition in patients with tuberculosis is a risk factor associated with early death. *Transactions of the Royal Society of Tropical Medicine and Hygiene*, 96(3): 291–294.

Intermezzo 4: Stuart Gillespie, Senior Research Fellow, International Food Policy Research Institute (IFPRI), and Director, Regional Network on AIDS, Livelihoods and Food Security (RENEWAL). 2007. "AIDS and Hunger: Challenges and Responses".

Gillespie, S. 2007. *AIDS and Hunger: Challenges and Responses.* Washington DC, IFPRI. Posted at: http://www.ifpri.org/renewal

Intermezzo 5: WHO Regional Office for the Eastern Mediterranean, 2007. "Food Support and the Treatment of Tuberculosis".

World Health Organization. 2007. The five elements of DOTS. Geneva. Posted at: http://www.who.int/tb/dots

Intermezzo 6: Dr Camila Corvalan, Emory University, University of Chile, 2007. "Nutrition transition in Latin America: The experience of the Chilean National Nursery School Council Program".

Albala, C., Vio, F., Kain, J. & Uauy, R. 2001. Nutrition transition in Latin America: The case of Chile. *Nutrition Reviews*, 59(6): 170–176.

De Onis, M. & Blosner, M. 2000. Prevalence and trends of overweight among preschool children in developing countries. *American Journal of Clinical Nutrition*, 72(4): 1032–1039.

Kain, J., Uauy, R., Lera, L., Taibo, M. & Albala, C. 2005. Trends in height and BMI of 6-year-old children during the nutrition transition in Chile. *Obesity Research*, 13(12): 2178–2186.

Monteiro, C.A., Moura, E.C., Conde, W.L. & Popkin, B.M. 2004. Socioeconomic status and obesity in adult populations of developing countries: a review. *Bulletin of the World Health Organization*, 82: 940–946.

Popkin, B.M. 1994. The nutrition transition in low-income countries: An emerging crisis. *Nutrition Reviews*, 52: 285–298.

Uauy, R., Albala, C. & Kain, J. 2001. Obesity trends in Latin America: Transiting from under- to overweight. *Journal of Nutrition*, 131(3): 893S–899S.

World Health Organization. 2003. Diet, nutrition and the prevention of chronic diseases. Report of a Joint WHO/FAO expert consultation. *Technical Report Series* No. 916.

Parts III and IV

14th Dalai Lama. (b. 1935). *Living a Life of Compassion.* Posted at: http://www.betterworldheroes.com/dalai-lama.htm

Alderman, H. & Behrman, J.R. 2004. *Estimated Economic Benefits of Reducing Low Birth Weight in Low-Income Countries.* Washington DC, World Bank.

Alderman, H., Appleton, S., Haddad, L., Song, L. & Yohannes, Y. 2001. *Reducing Child Malnutrition: How Far Does Income Growth Take Us?* Nottingham, UK, University of Nottingham Centre for Research in Economic Development and International Trade.

Barro, R.J. 1990. Government spending in a simple model of endogenous growth. *Journal of Political Economy*, 98(5): 103–126.

Behrman, J.R., Alderman, H. & Hoddinott, J. 2004. Hunger and malnutrition. *In* B. Lomborg. ed. 2004. *Global Crises, Global Solutions*. Cambridge, UK, Cambridge University Press.

Binswanger, H.P. & Landell-Mills, P. 1995. *The World Bank's Strategy for Reducing Poverty and Hunger: A Report to the Development Community*. Washington DC, World Bank.

Bloom, D.E., Canning, D. & Sevilla, J. 2001. The effect of health on economic growth: Theory and evidence. *NBER Working Paper* No. 8587. Cambridge, MA, National Bureau of Economic Research.

Brooker, S., Kabatereine, N.B., Fleming, F. & Devlin, N. 2007. *Cost and Cost-Effectiveness of Nationwide School-Based Helminth Control in Uganda: Intra-Country Variation and Effects of Scaling-Up*. London, London School of Hygiene and Tropical Medicine. Forthcoming.

Buse, K. & Harmer A. 2007. Global health: Making partnerships work. *ODI Briefing Paper*, 15 January: 1–4.

Carroll, L. (1832–1898). *Alice's Adventures in Wonderland*. Posted at: http://quotes.zaadz.com/Lewis_Carroll

Chatterjee, M. & Measham, A.R. 1999. *Wasting Away. The Crisis of Malnutrition in India*. Washington DC, World Bank.

Chekhov, A. (1860–1904). *Five Plays*. Posted at: http://thinkexist.com/quotation/knowledge_is_of_no_value_unless_you_put_it_into/221938.html

Chowdhury, A.M., Karim, F., Sarkar, S.K., Cash R.A. & Bhuiya, A. 1997. The status of ORT in Bangladesh: How widely is it used? *Health Policy and Planning*, 12(1): 58–66.

Disease Control Priorities Project. 2007. *Water, Sanitation, and Hygiene: Simple, Effective Solutions Save Lives*. Washington DC, The World Bank Group. Posted at: http://www.dcp2.org/file/81/DCPP-Water.pdf

Easterly, W.R. 2006. *The White Man's Burden*. New York, Penguin Books.

Economic Commission for Latin America and the Caribbean & World Food Programme. 2007. *Hacia la Erradicación de la Desnutrición Crónica Infantil en Centroamérica y Republicans Dominicana*. PowerPoint presentation given at the Regional Technical Consultation on Monitoring and Evaluation, Panama; *Busca Centroamérica Erradicar Desnutrición Infantile*. Panama. Posted at: http://www.milenio.com/index.php/2007/02/01/37334/

Edison, T. (1847–1931). Posted at: http://www.spice-of-life.com/quotes.html

Edwards, S. 1998. Openness, productivity and growth: What do we really know? *Economic Journal*, 108(447): 383–398.

Fernholz, R., Fernholz, F., Jayasekera, C. & Rueda, F. 2007. Practical interventions: Political choices. Technical paper for the *World Hunger Series 2007*. Durham, NC, Duke University.

Hall, R. & Jones, C. 1998. Why do some countries produce so much more output per worker than others? *Department of Economics Working Paper*. Palo Alto, CA, Stanford University.

Hawley, W.A., Ter Kuile, F.O., Steketee, R.S., Nahlen, B.L., Terlouw, D.J., Gimnig, J.E., Shi, Y.P., Vulule, J.M., Alaii, J.A., Hightower, A.W., Kolczak, M.S., Kariuki, S.K. & Phillips-Howard, P.A. 2003. Implications of the Western Kenya permethrin–treated bed net study for policy, program implementation and future research. *American Journal of Tropical Medicine and Hygiene*, 68(4) (Supplement): 168–173.

Horton, S. 1999. Opportunities for investments in nutrition in low-income Asia. *Asian Development Review* 17(1, 2): 246–273.

Horton, S. & Ross, J. 2003. The Economics of Iron Deficiency. *Food Policy*, 28(1): 51–75.

Komlos, J. & Lauderdale, B.E. 2007. Underperformance in affluence: The remarkable relative decline in U.S. heights in the second half of the 20th century. *Social Science Quarterly*, 88(2): 283–303.

Migele, J., Ombeki, S., Ayalo, M., Biggerstaff, M. & Quick, R. 2007. Diarrhoea prevention in a Kenyan school through the use of a simple safe water and hygiene intervention. *American Journal of Tropical Medicine and Hygiene*, 76(2): 351–353.

Organisation for Economic Co-operation and Development. 2007. *Statistics Portal.* Paris. Posted at: http://www.oecd.org/statsportal/0,3352,en_2825_293564_1_1_1_1_1,00.html

Osborne, D. & Gaebler, T. 1992. Reinventing government: How the entrepreneurial spirit is transforming the public sector. *In* J.Z. Kusek and R.C. Rist. 2004. *Ten Steps to a Results-Based Monitoring and Evaluation System. A Handbook for Development Practitioners.* Washington DC, World Bank.

Sanghvi, T., Ross, J. & Heyman, H. 2007. Why is reducing vitamin and mineral deficiencies critical for development? The links between VMDs and survival, health, education and productivity. *Food and Nutrition Bulletin*, 28 (1) (Supplement): S170.

Save The Children UK. 2007. *Addressing Chronic Food Insecurity through Productive Safety Nets.* Contribution to the *World Hunger Series 2007.* Rome, WFP.

Shiffman, J. 2006. Donor funding priorities for communicable disease control in the developing world. *Health Policy and Planning*, 21(6): 411– 420.

Shrimpton, R., Victoria, R., De Onis, M., Rosangela, C.L., Bloessner, M. & Clugstong. G. 2001. Worldwide timing of growth faltering: Implications for nutritional interventions. *Pediatrics*, 107: 1–7.

Skoufias, E. & Parker, S.W. 2001. Conditional cash transfers and their impact on child work and schooling: Evidence from the PROGRESA Program in Mexico. *Discussion Paper No. 123.* Washington DC, IFPRI.

Tanner, J.M. 1986. Growth as a mirror of the condition of society: Secular trends and class distinctions. *In* Demirjian, A. ed. 1986. *Human Growth: A Multidisciplinary Review.* London, Taylor and Francis Ltd.

Ter Kuile, F.O., Turlouw, D.J., Phillips-Howard, P.A., Hawley, W.A., Friedman, J.F., Kariuki, S.K., Shi, Y.P., Kolczak, M.S., Lal, V.A., Vulule, J.M. & Nahlen B.L. 2003. Reduction of malaria during pregnancy by permethrin-treated bed nets in an area of intense perennial malaria transmission in western Kenya. *American Journal of Tropical Medicine and Hygiene*, 68(4) (Supplement): 50–60.

Thomas, D. & Strauss, J. 1997. Health and wages: Evidence on men and women in urban Brazil. *Journal of Econometrics*, 77: 159–185.

United Nations Children's Fund. 2007a. *Infant and Young Child Feeding and Care. Protecting, Promoting and Supporting Breastfeeding.* New York. Posted at: http://www.unicef.org/nutrition/index_breastfeeding.html
___ 2007b. *Immunization. The Challenge.* New York. Posted at: http://www.childinfo.org/areas/immunization/
___ 2007c. 4.5 million children across Ghana to be de-wormed. *UNICEF News. New York.* Posted at: http://www.unicef.org/media/media_38248.html

United States Department of Agriculture. 2005. *Making It Happen! School Nutrition Success Stories.* Washington DC, Food and Nutrition Services, USDA.

World Bank. 2004. *World Development Report 2004: Making Services Work for Poor People.* Washington DC, World Bank.
___ 2006. *Repositioning Nutrition as Central to Development: A Strategy for Large-Scale Action.* Washington DC, World Bank.

World Food Programme. 2000. *Food and Nutrition Handbook.* Rome, WFP.
___ 2006a. *HIV/AIDS and Nutrition. Food In The Fight Against AIDS. Facts and Figures Worksheet.* Rome, WFP.
___ 2006b. *Afghanistan De-Worming Campaign.* Rome, WFP.
___ 2007. *Annual Performance Report for 2006.* Rome, WFP.

World Health Organization. 2006. *The World Health Report: Working Together for Health.* Geneva. Posted at: http://www.who.int/whr/2006/en
___ 2007. *World Health Statistics 2007.* Geneva, WHO.

World Health Organization & United Nations Children's Fund. 2006. *Oral Rehydration Salts – Production of the New ORS.* Geneva. *Posted at:* http://www.who.int/child-adolescent-health/New_Publications/CHILD_HEALTH/WHO_FCH_CAH_06.1.pdf

Intermezzo 7: Dr Stanley Zlotkin, Professor, Nutritional Sciences and Public Health Sciences, University of Toronto. 2007. "Sprinkles: An innovative, cost-effective approach to providing micronutrients for children".

Agostoni, C., Giovannini, M., Sala, D., Usuelli, M.L., Francescato, G., Braga, M., Riva, E., Martiello, A.C., Marangoni, F. & Galli, C. 2006. Double-blind, placebo-controlled trial comparing effects of supplementation with two different combinations of micronutrients delivered as Sprinkles on growth, anaemia, and iron deficiency in Cambodian infants. *Journal of Paediatric Gastroenterology and Nutrition*, 42: 306–12.

De Pee, S., Moench-Pfanner, R., Martini, E., Zlotkin, S., Darton-Hill, I. & Bloem, M.W. 2006. Home-fortification in emergency response and transition programming: Experiences in Aceh and Nias, Indonesia. *Food Nutrition Bulletin* (forthcoming).

Hirve, S., Bhave, S., Bavdekar, A., Naik, S., Pandit, A., Schauer, C., Christofides, A., Hyder, Z. & Zlotkin, S. 2007. Low dose of Sprinkles: An innovative approach to treat iron deficiency anaemia in infants and young children. *Indian Paediatrics*, 44: 91–100.

Menon, P., Ruel, M.T., Loechl, C.U., Arimond, M., Habicht, J-P., Pelto, G. & Michaud, L. 2007. Micronutrient Sprinkles reduce anaemia among 9- to 24-month-old children when delivered through an integrated health and nutrition program in rural Haiti. *Journal of Nutrition*, 137: 1023–1030.

World Health Organization. 2006. *Preventing and Controlling Micronutrient Deficiencies in Populations Affected by an Emergency*. Joint Statement by WHO/WFP/UNICEF. Geneva. Posted at: http://www.who.int/nutrition/publications/WHO_WFP_UNICEFstatement.pdf

Zlotkin, S., Arthur, P., Antwi, K.Y. & Yeung, G. 2001. Treatment of anaemia with microencapsulated ferrous fumarate plus ascorbic acid supplied as Sprinkles to complementary (weaning) foods. *American Journal of Clinical Nutrition*, 74:791–795.

Intermezzo 8: WFP Latin America and the Caribbean Regional Bureau. 2007. "Partnerships to overcome child undernutrition in Latin America and the Caribbean".

Economic Commission for Latin America and the Caribbean & World Food Programme. 2007. *Hacia la Erradicación de la Desnutrición Crónica Infantil en Centroamérica y Republicans Dominicana*. PowerPoint presentation given at the Regional Technical Consultation on Monitoring and Evaluation, Panama; *Busca Centroamérica Erradicar Desnutrición Infantile*. Panama. Posted at: http://www.milenio.com/index.php/2007/02/01/37334/

Inter-American Development Bank. 2007. *Acuerdo Sobre Desnutrición Infantil es Paso Clave en Lucha contra la Pobreza, Afirman BID y PMA*. IDB Press Release. Posted at: http://www.iadb.org/news/articledetail.cfm

Walsh, E.A. 2007. *Political Database of the Americas*. Washington DC, Centre for Latin American Studies, Georgetown University. Posted at: http://www12.georgetown.edu/sfs/clas/

Intermezzo 9: Valid International. 2007. "From Research to Action".

Bahwere, P., Guerrero, S., Sadler, K. & Collins, S. 2005. *Study to Examine the Use of CTC as an Entry Point for the Support to HIV Affected in Malawi*. Oxford, UK, Valid International.

Bahwere, P., Joshua, M.C., Sadler, K., Tanner, C., Piwoz, E., Guerrero, S. & Collins, S. 2005. *Integrating HIV Services into a Community-Based Therapeutic Care (CTC) Programme in Malawi: An Operational Research Study*. Unpublished.

Bahwere, P., Sadler, K. & Collins, S. 2005. *The treatment of Severely Malnourished HIV Positive Adults Using Ready to Use Therapeutic Food in Home Based Care*. Unpublished.

Brewster, D.R., Manary, M.J. & Graham, S.M. 1997. Case management of kwashiorkor: An intervention project at seven nutrition rehabilitation centres in Malawi. *European Journal of Clinical Nutrition*, 51(3): 139–147.

Brewster, D.R. 2004. Improving quality of care for severe malnutrition. *Lancet*, 363(9426): 2088–2089.

Collins, S. 2004. Community-based therapeutic care: A new paradigm for selective feeding in nutritional crises. *Humanitarian Policy Network paper*. No. 48. London, Overseas Development Institute.

Collins, S., Dent, N., Binns, P., Bahwere, P., Sadler, K. & Hallam, A. 2006. Management of severe acute malnutrition in children. *Lancet*, 368(9551): 1992–2000.

Collins, S., Sadler, K., Dent, N., Khara, T., Guerrero, S., Myatt, M., Saboya, M. & Walsh, A. 2006. Key issues in the success of community-based management of severe malnutrition. *Food Nutrition Bulletin*, 27(3): S49–S82.

Guerrero, S., Bahwere, P., Sadler, K. & Collins, S. 2005. Integrating CTC and HIV/AIDS support in Malawi. *Field Exchange*, 25: 8–10.

Kessler, L., Daley, H., Malenga, G. & Graham, S. 2000. The impact of the human immunodeficiency virus type 1 on the management of severe malnutrition in Malawi. *Annals of Tropical Paediatrics*, 20(1): 50–56.

National Statistical Office, Malawi. 2005. *Demographic and Health Survey 2004 and 2005*. Calverton, MD, ORC Macro.

Sadler, K., Myatt, M., Feleke, T. & Collins, S. 2007. A comparison of the programme coverage of two therapeutic feeding interventions implemented in neighbouring districts of Malawi. *Public Health Nutrition*, 10: 907–913.

Intermezzo 10: Rosemary Fernholz and Channa Jayasekera, Duke University, 2007. "Nutrition a Priority in Thailand".

Tontisirin, K., Kachondham, Y. & Winichagoon, P. 1992. Trends in the development of Thailand's nutrition and health plans and programs. *Asia Pacific Journal of Clinical Nutrition*, 1: 231–238.

Text notes

1 Life expectancy for 2002 from WHO (2007e). Stature for 2002 from CDC (2004), averaged for the 30–40 age group.

2 The wording of the quotation has been modified with the permission of the author.

3 The quotation has been paraphrased. The original is: *Hunger leads first, it is true, to the concentration of the whole energy of the body on the problem of getting food. Every thought and emotion of the starving man is fixed on this one primary need. But if he fails to obtain it, there are no complex psychoses for observation, but merely the gradual lowering of the whole vitality of the body, and the lethargy which leads to death.*

4 Not all countries have data for both indicators – undernourishment and underweight – for both periods (1990–2003 for undernourishment and 1990 and 1997–2006 for underweight). The table at the bottom of the page shows the number of developing countries and LIFDCs for which data are available.
Data for undernourishment come from FAO (2006). The WHO regional classification was used to derive regional breakdowns. Regional prevalence is calculated from the number of undernourished people in each region divided by the total population of the countries considered.

5 For a description of the method used to calculate progress towards reaching the MDGs, see the Resource Compendium, Table 10.
Of the 103 developing and transition countries analysed, the 15 poorest countries based on GNI are: Burundi, Democratic Republic of the Congo, Eritrea, Ethiopia, The Gambia, Guinea-Bissau, Liberia, Madagascar, Malawi, Mozambique, Nepal, Niger, Rwanda, Sierra Leone, Uganda.
The 15 richest developing and transition countries in terms of GNI are: Antigua and Barbuda, Botswana, Chile, Croatia, Gabon, Lebanon, Libya, Mauritius, Mexico, Palau, St Kitts-Nevis, Saudi Arabia, the Seychelles, South Africa, Trinidad and Tobago.

6 Excess mortality is operationally defined as a crude mortality rate exceeding 1 death per 10,000 people per day (WFP, 2004).

7 The Des Moines Declaration is a call by the Laureates, Founders and Council of Advisers of the World Food Prize Organization for accelerated action against hunger (Des Moines Declaration, 2004).

8 Data from demographic and health surveys conducted in Cambodia (2000), Bangladesh (2004), Chad (2004), United Republic of Tanzania (2004) and Ethiopia (2005). Method follows R. Shrimpton et al. (2001).

9 Komlos and Lauderdale paraphrased the quotation from James M. Tanner (1986): "Growth is a mirror of the condition of society".

10 Valid International is a research and consultancy company that was formed in 1999 to try to improve the impact of humanitarian action through action-oriented research. Over the last five years, a partnership between Valid International and Concern Worldwide – an Irish relief and development organization – has pioneered CTC, which is a new approach for managing the severely acute malnourished at home using social mobilization techniques and energy-dense, ready-to-use therapeutic food enriched with minerals and vitamins.

11 The quotation has been modified: the word "mankind" was changed to "humankind".

Number of Developing Countries

	Both indicators		Undernourishment		Underweight	
	Developing countries	Of which are LIFDCs	Developing countries	Of which are LIFDCs	Developing countries	Of which are LIFDCs
Africa	28	27	38	34	28	27
Eastern Mediterranean	8	6	10	6	8	6
Europe	3	3	17	9	3	3
LAC	9	4	9	4	9	4
South East Asia	7	4	8	6	7	5
Western Pacific	6	5	6	5	6	5

Costing the essential solutions

Food and cash-based transfers

Estimated cost of one adequate ration for a person entirely reliant on food assistance is based on 400 g cereal (rice), 60 g pulses (beans), 25 g vegetable oil, 50 g fortified blended food (wheat–soya blend), 15 g sugar and 5 g iodized salt.

Sources:

- World Food Programme. 2000. *Food and Nutrition Handbook*. Rome, WFP.
- Revised F.O.B. prices for WFP-supplied commodities 15 May 2006 CFO2006/002.

Targeted micronutrient supplementation

Iron: supplementation is provided weekly for schoolchildren over an extended period.

Source:

- Horton, S. 2006. The economics of food fortification. *Journal of Nutrition* 136:1068–1071.

Iron + folic acid: tablets are provided daily for one year covering pregnancy and initial lactation. The cost is based on 1998 UNICEF prices for providing 60 mg iron + 400 mg folic acid. Providing tablets for one year costs US$9.85 per woman.

Source:

- Gillespie, D., Karklins, S., Creanga, A., Khan, S. & Cho, N. 2007. *Scaling Up Health Technologies*. Report prepared for the Bill and Melinda Gates Foundation. Baltimore, MD, Bloomberg School of Public Health, Johns Hopkins University.

Vitamin A: supplementation is provided through a community-based distribution system.

Source:

- World Bank. 2004. *Vitamin A at a Glance*. Washington DC, World Bank.

Zinc: each treatment includes 14 tablets.

Source:

- Gillespie, D., Karklins, S., Creanga, A., Khan, S. & Cho, N. 2007. *Scaling Up Health Technologies*. Report prepared for the Bill and Melinda Gates Foundation. Baltimore, MD, Bloomberg School of Public Health, Johns Hopkins University.

Supplementary feeding

The on-site ration is based on 125 g corn–soya blend, 20 g vegetable oil, 30 g pulses (beans). A take-home ration is based on 250 g corn–soya blend, 25 g vegetable oil and 20 g sugar.

Sources:

- World Food Programme. 2002. *WFP Emergency Field Operations Pocketbook*. Rome, WFP.
- Revised F.O.B. prices for WFP-supplied commodities 15 May 2006 CFO2006/002.

Complementary feeding

One ration is based on 250 g corn–soya blend, 25 g vegetable oil and 20 g sugar.

Sources:

- World Food Programme. 2002. *WFP Emergency Field Operations Pocketbook*. Rome, WFP.
- Revised F.O.B. prices for WFP-supplied commodities 15 May 2006 CFO2006/002.

Large-scale food fortification

Iodine, iron and vitamin A: cost per person per year of fortification is estimated for a hypothetical large, low-income developing country.

Zinc: the cost assumes a daily intake of 150 g per person.

Source:

- World Health Organization and Food and Agriculture Organization of the United Nations. 2006. *Guidelines on Food Fortification with Micronutrients*. Geneva, WHO.

Home fortification

Supplementation is based on the cost of Sprinkles. The typical micronutrient mix is iron, zinc, iodine, vitamins A, C and D, and folic acid. Costs vary according to the nutritional status of beneficiaries, the micronutrient mix and delivery and distribution costs.

Source:

- Zlotkin, S. 2007. Sprinkles: An innovative, cost-effective approach to provide micronutrients to children. Technical paper for the *World Hunger Series 2007*.

Nutritional support for disease treatment (HIV/AIDS and TB)

A standard support package, for one month, including 333 g corn–soya blend, 33 g oil and 167 g maize meal.

Sources:

- World Food Programme. 2006. Cost of Nutritional Support for HIV/AIDS Projects, Rome, WFP.
- Revised F.O.B. prices for WFP-supplied commodities 15 May 2006 CFO2006/002.

Infectious disease prevention

Anti-malaria drugs and bed nets

Source:

- Gillespie, D., Karklins, S., Creanga, A., Khan, S. & Cho, N. 2007. *Scaling Up Health Technologies*. Report prepared for the Bill and Melinda Gates Foundation. Baltimore, MD, Bloomberg School of Public Health, Johns Hopkins University.

DOTS: community-based DOTS treatment costs US$128 per patient and US$203 for a patient treated through a health facility-based DOTS programme.

Source:

- Gillespie, D., Karklins, S., Creanga, A., Khan, S. & Cho, N. 2007. *Scaling Up Health Technologies*. Report prepared for the Bill and Melinda Gates Foundation. Baltimore, MD, Bloomberg School of Public Health, Johns Hopkins University.

Deworming tablets and delivery

Source:

- World Food Programme. 2007. *Food for Education Works. A Review of WFP FFE Programme Monitoring and Evaluation 2002–2006*. Rome, WFP.

ORT

Source:

- Gillespie, D., Karklins, S., Creanga, A., Khan, S. & Cho, N. 2007. *Scaling Up Health Technologies*. Report prepared for the Bill and Melinda Gates Foundation. Baltimore, MD, Bloomberg School of Public Health, Johns Hopkins University.

Male condoms: calculation is based on average costs for bulk purchase of male condoms.

Source:

- United Nations Population Fund. 2005. *Condom Programming for HIV Prevention*. New York, UNFPA.

Diversify diets and promote quality food consumption

School meals: a more substantial food ration of 150 g cereal, 30 g pulses, 5 g oil and 4 g salt (694 kcal) is provided at school, and eaten during school hours. The average cost of one school meal per beneficiary is US$20 per year and US$0.10 per day.

School gardens as a part of the full essential package: for school feeding, the package includes deworming, micronutrient supplementation, improved stoves, water and sanitation at school, health education, HIV/AIDS education, psychosocial support and malarial prevention. The average cost of this support per beneficiary is US$16 per year.

Source:

- World Food Programme. 2007. Standardization of selected Food for Education terms and figures for advocacy purposes. Internal document. Rome, WFP.

Transfer of knowledge on healthcare and food practices

Breastfeeding promotion: average costs from community-based assessments in sub-Saharan Africa vary according to the components of this activity.

Community-based nutrition programmes: costs include promotion of breastfeeding, counselling and education on optimal child feeding, prevention of diarrhoeal disease and growth monitoring.

Source:

- Caulfield, L.E., Richard, S.A., Rivera, J.A., Musgrove, P. & Black, R.E. 2006. Stunting, wasting, and micronutrient deficiency disorders. *Disease Control Priorities in Developing Countries*. New York, Oxford University Press.

Deworming training for teachers: this example is from Madagascar, where the health-education curriculum is supported by community information, education, and communication (IEC) with twice-yearly deworming and iron folate delivered by teachers. Over three years, 14,000 teachers and 430,000 students were trained in 4,585 schools, costing between US$0.78 and US$1.08 per capita per year.

Source:

- Disease Control Priorities Project. 2006. *Nine Low- and Middle-Income Countries and How They Use FRESH*. Washington DC, The World Bank Group. Posted at: http://www.dcp2.org/pubs/DCP/58/Table/58.4.

HIV education: costs include awareness education for children and training for parents and teachers in a school feeding programme.

Source:

- World Food Programme. 2006. *Standard Essential Package for School Feeding*. Unpublished. Rome, WFP.

Nutrition education for pregnant women: this is based on 1985 costs updated to reflect 2006 purchasing power.

Source:

- World Bank. 1994. *Enriching Lives Overcoming Vitamin and Mineral Malnutrition in Developing Countries.* Washington DC, World Bank.

Access to clean water and improved sanitation

Clean water: the cost depends on the number of units included in a system. For example, a system for four households would cost US$525 per unit; a larger system with 500–1,000 households would cost US$210 per unit.

Source:

- US Environmental Protection Agency. 2007. Cost of evaluation of Point-of-Use and Point-of-Entry Treatment Units for Small Systems: Cost Estimating Tool and User Guide. Washington DC, EAP.

Improved sanitation

Source:

- EU Water Initiative Eastern Europe, Caucasus and Central Asia Working Group. 2006. *Document 14: Rural Water Supply and Sanitation: technology overview and cost functions.* Prepared for the Environmental Action Programme for Central and Eastern Europe (EAP) Task Force Group of Senior Officials on the Reforms of the Water Supply and Sanitation Sector. Paris, OECD.

Methodology for maps

Map data and methods

Country boundaries

All map boundaries used in this publication are based on FAO GAUL – Global Administrative Unit Layer http://www.fao.org/geonetwork/srv/en/metadata.show?id=12691

Map projection

All maps present in this publication are in Robinson projection, datum WGS 84.

Scaling

Scaling for the underweight and mortality for children under 5 indicators is based on the international standard as described in WFP and CDC. 2005. *A Manual: Measuring and Interpreting Malnutrition and Mortality. Rome.* Rome, WFP.

Map construction

Map A – Hunger and health across the world

Data source:

- Underweight: World Health Organization. 2007. *World Health Statistics 2007*. Geneva, WHO.

Map 1 – Hidden hunger across the world

The map shows iron and vitamin A deficiencies as a proxy for micronutrient deficiencies among children under 5.

- Iron deficiency reflects the prevalence of detected cases of iron deficiency among children under 5.
- The classification of iron deficiencies in children under 5 is based on the international standard.
- The scaling of underweight rates for children under 5 is equal to or greater than 20 percent.

Data sources:

- Underweight: World Health Organization. 2007. *World Health Statistics 2007*. Geneva, WHO.
- Micronutrient deficiency: The Micronutrient Initiative and United Nations Children's Fund. 2004. *Vitamin and Mineral Deficiency: A Global Progress Report*. New York, UNICEF.

Map 2 – Hunger and natural disasters

The map highlights countries with underweight rates for children under 5 equal to or greater than 20 percent. The indicator for total number of affected people is the sum of people affected by natural disasters that occurred from 2000 to 2007.

Data sources:

- Natural disasters: Centre for Research on Epidemiology of Disasters. 2007. EM_DAT: The OFDA/CRED International Disaster Database. Université Catholique de Louvain. Posted at: http://www.em-dat.net
- Underweight: World Health Organization. 2007. *World Health Statistics 2007*. Geneva, WHO.

Map 3 – Inequality of Hunger Across the World

The map highlights countries with underweight rates for children under 5 equal to or greater than 20 percent, wasting rates equal to or greater than 15 percent; a range of stunting rates is presented.

Data sources:

- Underweight and Stunting: World Health Organization. 2007. *World Health Statistics 2007*.Geneva, WHO.
- Wasting: United Nations Children's Fund. 2006. *The State of the World's Children 2007 – Women and Children: The Double Dividend of Gender Equality*. New York, UNICEF.

Map 4 – Mortality and childhood diseases

The map highlights countries with underweight rates for children under 5 equal to or greater than 20 percent. Under 5 mortality rates are presented for acute respiratory infections, diarrhoeal disease and malaria, with cut-offs based on WHO standards.

Data source:

- World Health Organization. 2007. *World Health Statistics 2007*. Geneva, WHO.

Map 5 – The burden of malaria across the world

Burden values are stated in 2004 US dollar values, as estimated by the World Bank.

Data sources:

- Gross National Income: Organisation for Economic Co-operation and Development. 2006. *Statistical Annex of the 2006 Development Co-operation Report*. Paris. Posted at: www.oecd.org/dac/stats/dac/dcrannex
- Mortality rate: World Health Organization. 2007. *World Health Statistics 2007*. Geneva, WHO.

Map 6 – HIV/AIDS mortality in children under 5

The map highlights countries with underweight rates for children under 5 equal to or greater than 20 percent. The mortality rates for children under 5 are based on WHO standards.

Data source:

- World Health Organization. 2007. *World Health Statistics 2007*. Geneva, WHO.

Map 7 – Health inequalities across the world

Gross national income (GNI) is presented in 2004 US dollar values. The cut-offs for density of healthcare workers are based on the methods cited by WHO (2004) in *Joint Learning Initiative (JLI) Strategy Report: Human Resources for Health Overcoming the Crisis*. Posted at: http://www.globalhealthtrust.org/report/appendix2.pdf

Data sources:

- Gross National Income: Organisation for Economic Co-operation and Development. 2006. *Statistical Annex of the 2006 Development Co-operation Report*. Paris. Posted at: www.oecd.org/dac/stats/dac/dcrannex
- World Health Organization. 2007. *World Health Statistics 2007*. Geneva, WHO.

Map 8 – National commitment to health

The classification of public health expenditure is based on the UNDP 2006 Human Development Report.

Data source:

- Public health expenditure as a percentage of GNI: World Health Organization. 2007. *World Health Statistics 2007*. Geneva, WHO.

Map B – Hunger and health across the world

The map highlights countries with more than 1,500 deaths from pregnancy-related causes, including deaths occurring within 42 days of the termination of pregnancy, per 100,000 births. This threshold is based on WHO, UNICEF and UNFPA. 2004. *Maternal Mortality in 2000: Estimates Developed by WHO, UNICEF and UNFPA*. Geneva.

Data source:

- Underweight: World Health Organization. 2007. *World Health Statistics 2007*. Geneva, WHO.

Map B – Hunger and health across the world

The boundaries and the designations used on this map do not imply any official endorsement or acceptance by the United Nations.
Map produced by WFP VAM.

Data source: WHO, 2007

Russian Federation
Kazakhstan
Mongolia
Georgia
Armenia
Azerbaijan
Uzbekistan
Kyrgyzstan
Turkmenistan
Tajikistan
China
Dem. People's Rep. of Korea
Rep. of Korea
Japan
Syrian Arab Republic
Iraq
Islamic Rep. of Iran
Afghanistan
Jammu and Kashmir
Jordan
Pakistan
Kuwait
Bahrain
Qatar
Saudi Arabia
United Arab Emirates
Nepal
Bhutan
India
Bangladesh
Myanmar
Viet Nam
Lao People Dem. Rep.
Hong Kong (China)
Macao (China)
Taiwan
Oman
Yemen
Thailand
Cambodia
Philippines
Northern Mariana Islands (USA)
Guam (USA)
Djibouti
Ethiopia
Somalia
Sri Lanka
Maldives
Palau
Federated States of Micronesia
Kiribati
Brunei Darussalam
Malaysia
Singapore
Nauru
Uganda
Kenya
Rwanda
Burundi
United Republic of Tanzania
Seychelles
Indonesia
Papua New Guinea
Solomon Is.
Timor-Leste
Cocos (Keeling) Islands (Australia)
Christmas (Australia)
Comoros
Malawi
Mozambique
Madagascar
Vanuatu
Mauritius
Réunion (Fr.)
New Caledonia (Fr.)
Australia
Swaziland
New Zealand
wfp vam
vulnerability analysis and mapping